Charles Mayer Wetherill

The Manufacture of Vinegar

Its theory and practice, with especial reference to the quick process

Charles Mayer Wetherill

The Manufacture of Vinegar
Its theory and practice, with especial reference to the quick process

ISBN/EAN: 9783337130473

Printed in Europe, USA, Canada, Australia, Japan

Cover: Foto ©berggeist007 / pixelio.de

More available books at **www.hansebooks.com**

THE MANUFACTURE OF VINEGAR:

ITS

THEORY AND PRACTICE,

WITH

ESPECIAL REFERENCE TO THE QUICK PROCESS.

BY

CHARLES M. WETHERILL, Ph. D., M. D.,

MEMBER AMERICAN PHILOSOPHICAL SOCIETY; ACADEMY NATURAL SCIENCES OF PHILADELPHIA; MEMBER INDIANA STATE MEDICAL SOCIETY, ETC.

PHILADELPHIA:
LINDSAY AND BLAKISTON.
1860.

HENRY B. ASHMEAD, BOOK AND JOB PRINTER,
Sansom Street above Eleventh.

TO

R. OGDEN DOREMUS, A. M., M. D.,

PROFESSOR CHEMISTRY NEW YORK MEDICAL COLLEGE; NEW YORK COLLEGE OF PHARMACY, ETC.

THIS WORK IS

DEDICATED

WITH THE ESTEEM OF

THE AUTHOR.

PREFACE.

This book was written to fill a void in the American literature of its subject. It is true, that the modern vinegar process is no longer novel, and that a great part of the contents of this work may be found scattered over the several treatises of pure and applied chemistry; but on the other hand, there is no concise reliable American work accessible to the inexperienced or practised vinegar manufacturer, in which he can find what information is needful for carrying out his process to the best advantage.

Vinegar making will continue to be an extensive and widely diffused manufacture, and its importance will increase as our country becomes more widely occupied. This fact is self-evident, since a large portion of the freight upon each barrel of vinegar is paid upon the universally to be had water.

Having felt the need of such a work in the course of my profession as an analytical and consulting chemist, I had almost decided to translate the last edition of Otto's "*Lehrbuch der Essig Fabrikation,*" (1857); but upon further reflection I thought that by omitting some, amplifying some, and adding other subjects, (in fact, by rewriting the book,) a work could be obtained which would be more acceptable to the American public. This statement is made to avoid the imputation of assuming unjustly a portion of the results of Otto's labors. The general division of the work, many of the tables, all of the wood cuts, except two, and the quantitative analysis of vinegar, are borrowed from Otto.

I am also indebted to Gottliebs' Chemische Technologie, the U. S. Dispensatory, Dr. Ure's Encyclopedia, Johnston's Chemistry of Common Life, Regnault's Cours de Chimie, Kopp's Gechichte der Chemié, and to some other works for information. While making these acknowledgments, I desire also to assume for myself some information arising from an acquaintance with a practical experience of the manufacture of vinegar.

In my method of treating the subject, some things may appear too scientific for the unlearned, and others too trite for the better informed. If the work has failed in this respect, it has happened through the endeavor to reconcile the subject to these two classes of persons, thereby obtaining a larger audience.

CHARLES M. WETHERILL.

LAFAYETTE, INDIANA,
   June 1, 1860.

TABLE OF CONTENTS.

INTRODUCTORY AND HISTORICAL.

Early knowledge of vinegar, 13. Properties of crystalizable acetic acid, 16. Wood vinegar, 17. History of the manufacture of vinegar from alcohol, 21. What limits factory sales of vinegar, 23. Future improvements in the flavor of vinegar from alcohol, 25. Division of the subject, 27.

PART I.

THEORETICAL.

CHAPTER I.

CHEMICAL PRINCIPLES.—Chemical nature of the elements concerned in the vinegar manufacture, 31. Laws of chemical combination, constancy of composition, 34. Law of proportion, 36. Illustrated, 38. Chemical symbols, 39. Atomic theory, 42. Practical example of this theory applied to alcohol, 46.

CHAPTER II.

SUGAR.—Connection between sugar and vinegar, 48. Table of the composition of the saccharoid bodies, 49. Cellulose, 51; its properties, 52; its transformation to dextrine and sugar, 53. Properties of starch, 54; its occurrence in the vegetable kingdom, 56; its preparation, 57; its chemical nature, 59. Diastase, 60; its action in vegetation, 61. The gums, 63. Dextrine—its preparation by sulphuric acid, 64; by diastase, 65; by heat, 66. The sugars, 66; the consumption of, 67. Milk sugar, 68. Varieties of sugar, 69. Cane sugar, 71. Fruit sugar, 75. Raisin sugar, 76. Simple method of analyzing sugars, 78. Manufacture of raisin sugar from starch, 81. What is meant by polarized light, 83. Action of sugar upon polarized light, upon which is founded a method for the analysis of sugar, 88.

CHAPTER III.

ALCOHOL.—Action of ferments upon sugar, 91. Composition of yeast, 92; Liebig's theory of fermentation, 92. Conditions for fermentation, 94. How much alcohol a given weight of sugar can yield, 97. Practical application of fermentation—wine, 99. Tables of the sweetness and acidity of wines, 100. Table of the alcoholic strength of different wines, 101. Fermentation of saccharine solutions by the addition of yeast, 104. The art of brewing, 106. Manufacture of malt, 108. Steeping, 110. Couching, 111. Drying malt, 113. Mashing, 115. Concentration of the wort, 118. Average per centage of malt extract from the different grains, 119. Saccharometer test for the strength of worts, 120. Table of the specific gravity of saccharometer degrees, 121. The boiling and refrigeration of worts, 122. Phenomena of fermentation, 123. Properties of alcoholic solutions, 126. Absolute alcohol, 127. Alcoholic strength of solutions tested by the dilatometer, 130; by the thermometer, 131; by the specific gravity, 132. Simple method of taking specific gravities, 133. Hydrometers, or areometers, 137. Beaumé's hydrometer table, 141, 142. Gay Lussac's volumeter, 143; saccharometer, 143; acetometer, 144; alcoholometer, 144, Proof spirits—English, New York, Ohio, 145. Method of using hydrometers, 145. Alcoholometer tables, corrections for temperature, 147, 148. Specific gravity table for obtaining the strength of alcoholic solutions, 150. Rule and tables for the transformation of alcoholic volumes per cent. to weights, per cent., and conversely, 153. The calculation for making definite mixtures of alcohol and water, 154. Alcohol tests for wine and beer, 156. Alcohol tests for wine and beer by the saccharometer, 157. Apparent and real attenuation of worts, 158. Balling's attenuation tables, 160, 161.

CHAPTER IV.

ACETIC ACID.—Liebig's theory of the vinegar process, 162. This process, illustrated by symbols and numerically, 163. Particulars to be observed in the vinegar manufacture— nature of the ferment—strength of alcohol in the vinegar mixture, 164. The limits of temperature, 165. Table of density of aqueous solutions of acetic acid, 166. Restricted use of the vinegar hydrometer, 168. Tables for the conversion of hydrated to anhydrous per cents., and conversely, 169, 170. Tests for adulterated vinegar, 170. Acetometry, 172. Chemical acetometry by dry alkalies— by crystalized carbonate of soda, 174; by dry carbonate of soda, 177; by dry carbonate of potassa, 177. Table of vinegar tests to use with foregoing processes, 178. Method by weighing alkaline carbonated solutions, 178. Four tables to facilitate this method, 180. Method by measuring alkaline carbonated solutions, 182. Method by ammonia with Otto's acetometer, 189. Preparation of the ammonia test solution, 193. Table of Tralles' alcoholometer degrees, converted to specific gravities, 194. Table to facilitate the preparation of Otto's test solution, 195. Another method of preparing this solution, 198. Balling vinegar tester, 202. The acid strength of vinegar, compared with the alcoholic strength of the mixture, 203. Tables to facilitate this calculation, 204.

PART II.

PRACTICAL.

CHAPTER I.

GENERAL DETAILS.—General principles of the vinegar manufacture, 209. The wash or mixture, 213. The water to be employed, with tests of its purity, 214. Methods of improving bad water, 216. Water filter, 217. The factory buildings, 218.

CHAPTER II.

THE SLOW PROCESS.—Example illustrating the difference between the slow and quick vinegar processes, 220. The household manufacture of vinegar, 221. An improved and simple method for household manufacture, 222. The factory method, 223. The practice at Orleans, 224. Arrangement for a factory by this process, 225. Modifications of the slow process, 229.

CHAPTER III.

THE QUICK PROCESS.—Boerhave's method, 233. Doebereiner's method, 234. Preparation of the platinum sponge, 235. Costliness of this method, 237. The modern process, 237. The apparatus required, 238. Manufacture of the beech wood shavings, 241. The generators, 242. Details of the operation—to set the generators in action, 247. Modifications of this operation, 249. Practical example of the manufacture, 252. Mode of working with three generators and six mixing tubs, 254. Points to be observed in the quick vinegar process, 257. The quantity of air entering the generators, and the changes it experiences, 258. The temperature of the vinegar room, and of the wash, 261. Method of warming the wash, 262. Otto's practice to avoid heating the acid washes, 264. To regulate the quantity of ferment in the generators, 267. The advantages of a periodical over a constant flow of the wash, 268. Hint for obtaining an improved constant flow, 272.

CHAPTER IV.

EXAMPLES OF THE PRACTICE OF THE BEST EUROPEAN FACTORIES. Table of German measures, 273. Vinegar process in Germany—(1) a factory in the Dutchy of Brunswick, 274; (2) one in the city of Brunswick, 275; (3) factory in Beuthen, 276; (4) Schultze's method, 277; (5) a highly prized recipé, 278; (6) factory in N., (7) in S., (8) in L., 282. (9) Method in some manufactories, 283. (10) Improvements upon the quick vinegar process in England and Germany, 283. Comparative advantages of large over small generators, 285. Furnace for effecting a current of air through generators, 287.

CHAPTER V.

CONCLUSION.—To improve the flavor and odor of vinegar, 289. To improve the color, 290. Clearing vinegar, 291. Recipes for the preparation of vegetable and aromatic vinegars. Three methods for four thieves vinegar. Tarragon vinegar. Vinaigre aux fines herbes. Vinaigre á la Ravigote. Mustard vinegar. Raspberry vinegar. Rose, orange blossom, néroli, bergamot, and clove vinegars. Fumigating vinegar. Aromatic vinegars. Crème de vinaigre. pp. 292–4.

VINEGAR MANUFACTURE.

INTRODUCTORY AND HISTORICAL.

VINEGAR is doubtless the first acid with which man became acquainted. As it generates so readily in dilute solutions of alcohol, which contain a ferment, it must have been used cotemporaneously with wine,* itself a very ancient beverage and formed spontaneously from the expressed juice of the grape. "Noah planted a vineyard," and "drank of the wine" to intoxication. The account is so circumstantial, that no doubt is left that the patriarch was well acquainted with the art of vinous fermentation; and he could not have kept wine without witnessing its transformation into vinegar. The grateful use of the grape in allaying thirst, must have led at an early time, to expressing its juice for the purpose of drinking. The fermentation of this juice and its transformation to an exhilarating drink—wine, must have been at once known, and almost simultaneously the change from wine to vinegar.

* Vinegar.—Vin aigre—sour wine. Essig Acetum. οξος from οξυς, sour.

The pleasant effects of wine led, no doubt, to the trial of vinegar; and its power in assuaging thirst, its culinary, and conservative properties must soon have been appreciated.

Some of the chemical properties of acetic acid were also known at an early day; as for example, its solvent action, with effervescence, upon carbonates. This appears from the following Proverb of Solomon, (chap. xxv. 20.) "As he that taketh away a garment in cold weather, and as *vinegar upon nitre*, so is he that singeth songs to a heavy heart."

To understand the simile, we must know that the word "*neter*," in the original, is improperly translated "*nitre*." Luther has rendered it by "*chalk*," which is more suitable, as that substance does effervesce with vinegar, which has not the slightest action upon nitre. But strictly, the ancients understood by "*neter*," the carbonated alkali either of potash or of soda. The commotion which vinegar produces when poured upon one of these carbonates, may well be taken to represent the effects of hilarity upon a heavy heart.

That the solvent effects of vinegar were understood by the ancients, is shown by the well known anecdote of Cleopatra, related by Pliny. To gain a wager that she would consume at a single meal, the value of a million sesterces, she dissolved pearls in vinegar which she drank. This is also shown by the equally well known, but exaggerated account by Livy and Plutarch, that Hannibal overcame the difficulties offered by the rocks to the passage of his army over the Alps, by dissolving them with vinegar. Admitting the exaggeration, or the explanation which some give, viz.: that Hannibal used the vinegar by way of stratagem, to incite his men to greater exertion by the belief that the difficulties of the path were diminished, the case nevertheless shows that the solvent action of vinegar upon certain substances, was well known at that period.

Vitruvius also states that rocks which cannot be attacked by either fire or iron, will yield when heated and wet with vinegar. The vinegar of the ancients was that of wine, namely, a dilute solution of acetic acid, containing certain soluble and odorous matter derived from the grape.

We are indebted to the much abused alchymists, for the first knowledge of its purification and concentration by distillation.

Geber, who flourished in the eighth century, gives us the earliest description of this method. Albucases in 1100, and Basilius Valentinus in the fifteenth century, wrote upon the same subject, and relate their experince as to the best means of effecting the concentration of vinegar by distillation.

Tachenius, (1666,) taught how to make a still stronger acetic acid, by the distillation of verdigris, which is an acetate of copper. Stahl in 1697, strengthened vinegar by freezing out some of its water in cold weather. In 1702, he taught the method of obtaining strong acetic acid by neutralizing vinegar by an alkali and distilling the acetate thus formed with oil of vitriol. The Count de Lauraguais, (1759,) and the Marquis de Courtenvaux, (1768,) showed that the most concentrated acetic acid obtained from verdigris, was capable of crystallization. Lowitz, (1789,) taught how pure but weak acetic acid might be strengthened, by passing it repeatedly over charcoal powder. Its may be thus deprived of so much of its water that it crystalizes by cold.

This crystalizable acetic acid is the strongest which it is possible to obtain. Durande, (1777,) gave to it the name, which it still bears, of glacial acetic acid. It is manufactured now as it was at the close of the last century, by the distillation of one of the acetates* with a mineral acid. The following are its properties. Its density at 60° Fah. is 1.063. Sixty parts by weight contain fifty-one parts of anhydrous, or dry acetic acid, and nine of water. This water cannot be removed from the acid, without destroying its acid properties. It may be *replaced* by a *base*, in which case a salt, an acetate results. Thus, thirty-one parts of soda will take the place of the nine parts of water and eighty-two parts of acetate of soda will result. Glacial acetic acid is a colorless, volatile and inflammable liquid, congealing at 40° Fah. and boiling at 248°. It possesses a pungent, refreshing

* Generally acetate soda in the United States.

smell, and corrosive properties to most vegetable and animal substances. It attracts water from the air, is miscible in all proportions with water, alcohol and ether, and dissolves camphor and several resins.

Having thus traced the knowledge of the ingredient which gives value to vinegar, from the earliest to the present time, let us now consider, also historically, the process of its manufacture.

If we except the brilliant results which Bertholet has obtained within the last two years, enabling him to form in small quantities *from their elements*, alcohol and several hundred allied substances, we have only two sources of acetic acid. These are, 1st, the destructive distillation of vegetable and animal substances, especially wood, and 2nd., alcohol.

1. Wood vinegar was known as early as the year 1648, for the celebrated Glauber describes it in a work published at that time, as arising from the distillation of all vegetable matter. This kind of vinegar is manufactured in our day, but as it possesses a disagreeable empyreumatic odor from which it is freed with difficulty, it is never used in the household. It is employed in the arts for purposes in which the smell is not objectionable, and for manufacting the salts of acetic acid, and from these the concentrated acid. It finds its way in some localities into vinegar employed for culinary purposes. In such cases a concentrated and purified wood vinegar, of from

forty to fifty per cent. acid strength, is added to raise the per centage of acetic acid in vinegar made from spirits by the quick process. I will dismiss the subject of wood vinegar with a brief description of the general principles of its manufacture.

Dry wood is subjected to heat in close iron cylinders, which are provided with a conduit tube and worm for condensing the volatile products, and placed either vertically or horizontally in appropriate furnaces. At first, water (which is rejected), comes over, then a liquid, which on standing separates into two strata. The lower stratum is wood tar, it is of an oily nature and contains several valuable products, the upper layer is of a watery nature and is more abundant in quantity than the tar. It contains the vinegar and is called crude pyroxilic, or pyroligneous acid. It is of a dark color and peculiar smoky odor. It is very acid and contains from eight to ten per cent. of acetic acid. During the distillation of the wood an abundance of combustible gas is formed, which is passed into the furnace to burn, thereby saving fuel. A fine quality of charcoal is removed from the cylinders at the close of the operation.

The crude pyroxilic acid is purified and strengthened in the following manner. Quicklime is slaked and made into a thin smooth paste with water, and used for completely neutralizing the acid. By this means acetate of lime is formed.

A portion of the resinous matter is thrown down by the lime, but another portion enters into combination with it, and dissolves in the solution of acetate of lime, imparting to it a deep brown color. The solution is cleared either by a filter or by standing, and then concentrated by evaporation to one half its volume. Enough hydrochloric acid* is then added to show a weak acid reaction by litmus paper, and the concentration continued to dryness. By the action of the hydrochloric acid, an additional portion of resinous matter separates, which rises as scum in the boiling liquid and may readily be removed. The hydrochloric acid decomposes also certain bodies which were in combination with the lime, yielding volatile products, which are dissipated during the evaporation of the solution. The dry residue is heated very carefully to remove as much as possible its volatile impurities. The result is a crude acetate of lime of a dirty brown color.

From this crude salt, a solution of acetic acid may readily be obtained by distillation with sufficient hydrochloric acid to saturate its lime.

One hundred parts of perfectly dry, pure acetate of lime, require 140 parts of commercial muriatic acid (of density 1.16), to saturate its lime, and on distillation with these proportions an acid is obtained containing 40 per cent. of anhydrous or dry acetic acid. But the *crude* acetate of lime in question, is impure and contains (beside resinous

---

* Muriatic acid of commerce.

matter), a portion of chloride of calcium, formed by the addition of hydrochloric acid to the vinegar in the first stage of its purification. It is, therefore, necessary to perform an experiment upon a sample of the crude acetate to ascertain how much hydrochloric acid is necessary to liberate its acetic acid. This is important, for an insufficiency of hydrochloric acid involves a loss of acetic acid, and too much hydrochloric acid is not only of useless expense, but yields a vinegar containing this objectionable acid.

When the vinegar is required of less strength than 40 per cent., a little water is added to the mixture of hydrochloric acid and acetate of lime. The following proportions are sometimes used. One hundred parts of the crude acetate of lime, 90 to 95 of muriatic acid of density 1.16, and 25 parts of water. These yield from 95 to 100 parts of vinegar of 31 per cent. dry acetic acid. In round numbers, 150 quarts of crude pyroxilic acid yield about 70 pounds of acetic acid of the above strength. This acid contains a small quantity of hydrochloric acid, from which it is freed by a second distillation over a little carbonate of soda. The resulting acid is colorless, but possesses a faint empyreumatic smell, of which it may be deprived by employing in this second distillation from 2 to 3 per cent. of bi-chromate of potassa instead of the carbonate of soda.

The stills used in this manufacture, should have heads of stone-ware and worms of glass.

For the final rectification, these should be of silver if the purest acid be required.

2. Vinegar from alcohol.

The remaining source of vinegar, viz., alcohol, has for the objects of this work a greater interest. Every vinegar used for household purposes has from the earliest period of its use, had its source in alcohol; but until the year 1822, this alcohol was always taken as found already prepared in fermented juice of fruits, as wine, cider, &c. In 1814, Berzelius discovered the correct composition of acetic acid, and in the same year, De Saussure performed the same service for alcohol. These facts afforded the proper stand-point for understanding the theory of the transformation of alcohol to acetic acid; and, accordingly, in 1822, after Prof. Dœbereiner, of Jena, had discovered that a weak solution of alcohol brought in contact with platinum black,* in the presence of air, was converted into acetic acid, he was enabled to give the theory of vinegar manufacture which is now adopted. Reserving the development of this theory for a future page, I need only say that the quick vinegar process, in which diluted alcohol is exposed to the air in contact with beech shavings and ready formed vinegar at an elevated temperature, takes its date from Dœbereiner's discovery, and the process proceeds upon exactly the same principles. This process is very appropriately named the *quick* vinegar manufacture, for the

* Very fine powder of metallic platinum.

sole difference between it and the ancient method of making vinegar by the exposure of wine, cider, &c., to the air, consists in the greater speed with which vinegar may be obtained.

Boerhave* at the commencement of the eighteenth century, invented a process for quickening the manufacture of vinegar from wine, which involves some of the principles of the modern quick process, and which at one time was very generally employed in France. At this period, however, the true theory of the vinegar manufacture was not understood.

We have seen how, in former times, vinegar, a substance of universal use and indispensable in every household, could only be made in localities where wine or the fermented juice of some appropriate fruit was to be obtained. It was necessarily an expensive condiment, for besides employing wine, a costly substance, not only must considerable capital be idle during the length of time required for its manufacture by the old process, but since it contains so small a quantity of acid, its transport is expensive, for on every hundred pounds, freight must be paid for the carriage of from 90 to 97½ pounds of water. On the other

* Although foreign to the subject in hand, I cannot refrain noticing an experiment of this distinguished man, to show his patience and zeal in exposing the errors and deceits of some of the alchymists. Boerhave kept mercury at a slightly elevated temperature, in an open vessel for *fifteen years*, and proved that it experienced thereby no material change! He also distilled the same portion of pure mercury five hundred times with the like result!!

hand, the modern quick process enables vinegar to be made in every locality. The amount of capital and knowledge requisite are very small. Spirits may be, at a comparatively cheap rate, transported to the central point of a community, where the vinegar may be manufactured and distributed within a circle, the extent of which will depend upon the knowledge and skill of the manufacturer enabling him to make cheaply; upon the cost of material, fuel and wages; the freight on shipping the vinegar; the demand for the article, and the amount of competition. If, as Otto advises, a very strong vinegar be made, containing from 10 to 15 per cent. acid, the circle of its sale may be much extended; for while its transport will cost the same as a weaker article, one barrel may be converted into several by the addition of sufficient water to bring it to the usual strength at the locality where it is used. Not only so, but the cost of barreling and storage will be less, and the vinegar can be kept with less danger of spoiling.

We may see from these considerations that vinegar can be, ought to be, and will be manufactured in almost every community in our country, and it is the object of this book to diminish the business of those who are still selling as a secret among us, a discovery that has been in successful practical operation here (and in Europe very publicly), for the past thirty years.

Vinegar as a condiment is valuable, not only

in proportion to the acid it contains, but also according to the flavor which it possesses by reason of certain sweet smelling ethers existing in it in small quantities, and derived from the fruit of which the fermented liquor is made. These are absent in vinegar made from chemically pure alcohol; some of them are present in vinegar made from alcohol which contains naturally a little fusel oil.

In no respect has chemistry made so rapid progress as in the knowledge we possess within the last few years of the ethers; some of the sweet smelling ones being identical with the aromatic principles of fruits or the bouquet of vines. At the English Crystal Palace Exhibition, some of these were deposited as articles of manufacture, and are now for sale in this country as secrets, at exorbitant prices. The celebrated Mulder has lately written to show to what extent they are used in Europe in the artificial manufacture of wines.

The aroma of the jargonelle pear is exactly that of acetate of amyle oxide; apple oil that of valerianate of amyle oxide; pine apple oil that of butyric ether,* (or better propylic ether); quince oil that of pelargonate of ethyle oxide; the flavor of whisky is due to acetate of the oxide of capryle.† These different flavors are developed by dissolving the respective etherial substances in six or more times their bulk of alcohol.

* May be made from butter. † From *rancid* butter.

The natural acidity of fermented drinks does not, in all cases, depend upon acetic acid derived from a portion of their alcohol. It is due to certain substances contained in the must or mash, or derived therefrom during the fermentation. Thus to acetic acid, (vinegar in this case,) is due the acidity of malt beer; to lactic acid that of milk beer and cider; to tartaric acid that of wine; and to citric acid that of wine made from imperfectly ripened grapes. Vinegar, however, appears in all of these beverages if the fermentation has been pushed too far. The improvement of most wines by age is due, in part, to a separation of the acid bitartrate of potash, which gradually deposits from the liquor. These acids appear in vinegar made from the respective beverages, and are not found (except of course acetic acid), in vinegar made from pure alcohol. I have made these statements for the purpose of directing attention to a branch of the vinegar manufacture which, to my knowledge, has not yet received the notice which it merits, but which must result to the pecuniary advantage of those engaged in it. I mean the preparation of a vinegar by the quick process, which shall resemble and perhaps equal in every respect the finest wine vinegar; to be sold bottled, and at an advanced price, for salads and other table use. I have no doubt that this problem may be readily solved with but little experiment. I am convinced that several of the ethers with which we are acquainted actually give flavor to some of our fruits and wines, and consequently are not at all injurious to health. When these ethers were first introduced to flavor new varieties of confectionery, I performed an experiment upon myself, for the sole purpose of testing their injurious properties, by eating within twelve consecutive hours one pound of the fruit drops flavored with the so-called essence of jargonelle pear, apple, banana and pine apple, without the slightest unpleasant effect.

The moral quality of this manufacture will depend upon the manner by which the article is presented to the public. If given as actually obtained from wine, it will, of course, be an unwarrantable deception.

Several years ago, a convention of wine growers was held in Germany, to consider the improvement of their industry. They had suffered for a few preceding years from bad wine, owing to seasons unfavorable for ripening the grape. Liebig, on the ground that the diminution of strength in their wine was due to an insufficiency of sugar in the grape, owing, of course, to the elements of the season which preside over the formation of sugar, advised them to add before fermentation, a small portion of saccharine matter to the must. This proposition they rejected with horror, as equivalent to an *adulteration*. But would it not have been better to have taken the advice and to have enjoyed wines of *good* quality, rather than to have been content with the *inferior* natural article? Can any one doubt if Hofmann is successful in his search after a method of obtaining quinine *artificially*, that the profession will administer it upon an equal footing with that extracted from bark.

Imagination is a potent tyrant over the "*senses*" of the public; but it appears that the right way to gain their approval of any article, is to act openly with them, proving that it is not injurious, but equally good with the article to which they have been accustomed, and at a lower price. Though it is an abhorrent doctrine, to be honest merely because it "*pays*," it is nevertheless true that honesty is the best policy.

In treating the subject of the quick vinegar process, I have, (like Otto,) divided it into two distinct parts, the "*theoretical*" and the "*practical.*" In the first part I have endeavored to set forth fully all that is required, for an unlearned man to understand the principles of the manufacture. In the second part will be found the method for carrying these principles into practical operation. I would here earnestly urge upon whoever desires to enter this manufacture, to make himself well acquainted with its theory, for that will be his surest anchor to enable him to ride out the storms of competition, as it will place him in the position to obtain the product at the lowest cost. Every vinegar generator is an individual, and its operation must be studied for itself, by testing the strength of the alcohol mixture and of the resulting vinegar, and by noting the time required for the transformation. We are only then able to calculate the cost of the manufacture.

PART I.

THEORETICAL.

CHAPTER I.

CHEMICAL PRINCIPLES.

BEFORE we can understand the theory of the change of alcohol to vinegar, we must become acquainted with the chemical nature of the bodies in question. They belong to the department of organic chemistry.

Without aiming at strictly scientific accuracy, it will be sufficient for our purpose to observe, that organic chemistry treats of substances found in or obtained from vegetable or animal bodies. The most of these substances are composed of different proportions of carbon, hydrogen and oxygen, or of carbon, hydrogen, nitrogen and oxygen. These are called elements, because they have not been decomposed. What is the nature of these elements in their free state, and what the laws according to which they unite with each other?

Carbon is well known by the name of charcoal and lampblack. In a pure and crystalized state it is graphite (the black lead of pencils) and the diamond. It burns by union with another element (oxygen) found in the air, and gives rise to a gas, carbonic acid. This gas may be detected in small quantities in the atmosphere; it is also found in the bottom of some wells, in caves and valleys near volcanoes, and in coal mines where it is called choke damp. Carbonic acid is so much heavier than air, that it may be poured from one vessel into another like water. It is destructive to animal life when breathed, and extinguishes instantly a flame immersed in it.

Nitrogen and oxygen are gases constituting the chief bulk of the atmosphere, which, neglecting the other gases existing in small quantities, is when dry, composed of four-fifths nitrogen and one-fifth oxygen by measure. These gases are not in a state of chemical combination in the air, but *mixed*, which condition enables life and many processes of the arts requiring oxygen, to be carried on. Air has been examined, taken from the plains of Egypt, from the summits of Mt. Blanc and Chimbarazo, as well as from an elevation of 21,000 feet reached by a balloon, and has always been found to consist of nitrogen and oxygen in the above mentioned proportions.

Oxygen is a tasteless, inodorous and colorless gas, heavier than air and not inflammable, but enabling combustible bodies to burn by uniting and forming chemical compounds with them. It may be obtained in a pure state by raising to a red heat nitre, which is a compound of nitrogen, oxygen and potash.

Since oxygen is heavier than air, nitrogen must of course be lighter. This gas is also destitute of color, taste or smell, and is neither combustible nor a supporter of combustion. Though not poisonous, it deprives of life an animal immersed in it and instantly extinguishes a candle, because it does not afford what life and flame require, viz., oxygen. It may be obtained from air contained in a close vessel by immersing a stick of phosphorous, which robs the air of its oxygen. Nitrogen, as found in the air, exhibits a weak affinity for the rest of the elements, and plays the part of diluting the oxygen of our atmosphere enabling animals to live in it. With a large proportion of oxygen in our air, every animal would die of inflammation, and not only would the iron the smith is forging take fire, but his anvil also. A spark would be sufficient to kindle a conflagration which would consume every combustible substance upon our earth.

Hydrogen, the remaining element we have to consider, is the lightest known substance. It is an inodorous, tasteless and colorless gas, and exists in chemical combination with oxygen in water, which contains *by weight*, one-ninth hydrogen and eight-ninths oxygen. It burns with a very pale blue flame, which is almost colorless. We have thus in water two gases, one the most inflammable of known substances, the other the best supporter of combustion. When these two gases in the free state are mixed in the above proportions and kindled, combustion ensues with the formation of water and evolution of the most intense heat. Upon this principle depends the oxyhydrogen blowpipe, (of which an excellent form has been invented by our countryman, the late Dr. Hare,) which affords the source of the greatest heat under the control of man.

From these few elements, chemically combined in different proportions or according to different methods, arises the vast number of substances derived from the animal and vegetable world, like the infinite number of beautiful forms produced in the kaleidescope by the varying relative position of a few fragments of colored glass. Let us now consider the laws of chemical combination so far as they relate to the subject in hand.

The first to be mentioned is the *constancy of composition of chemical compounds*. It is by this law that analysis is useful to us, for it affirms that a given body is an *individual*, and consequently may be recognized, because it has always the same composition. The apparent exceptions to the above results of analysis, as when *different* bodies seem to have the *same* composition, are reconciled by supposing that the bodies in question have their elements put together according to a different order or method. As these exceptions do not concern the theory of the vinegar process, we will turn our attention to the law itself. Take water for an example, every 9 pounds of which contain 8 pounds of oxygen and 1 pound of hydrogen. We may freeze or vaporize water an infinite number of times, decompose it into its elements, recombine them to water, using if you please, hydrogen and oxygen taken from the antipodes. We may analyze the water found in the interior crevice of a quartz crystal, or wherever else it may occur upon the globe, and it will present the invariable composition, "in 9 weights of water, 8 weights of oxygen and 1 of hydrogen."

Whenever hydrogen and oxygen come together under the circumstances favorable to form water, they will unite in these porportions with the result, water. What is true of water holds good with every other chemical compound. Twenty-two pounds of carbonic acid always contain 6 pounds of carbon and 16 pounds of oxygen. One hundred and eighty pounds of sugar invariably consist of 72 pounds carbon, 12 hydrogen and 96 oxygen. Forty-six pounds of alcohol contain 24 pounds carbon, 6 pounds hydrogen and 16 pounds oxygen. In 60 pounds of acetic acid we will never find more or less than 4 pounds of carbon, 40 pounds of hydrogen and 32 pounds of oxygen.

Now, why should there be this constancy of composition in the same body, and how is it that a difference in the relative proportions of carbon, hydrogen and oxygen gives rise to bodies of totally different properties? These questions were only asked and answered towards the close of the last and commencement of the present century. Their consideration leads us to the second law of chemical combination, i. e., by *proportion*, which law is explained by the atomic theory. The law is very simple. It is derived from the analysis of chemical compounds. It is evident that we may state the result of analyses in different ways. I have given one way in the example of water, to wit, 9 parts by weight of water contain 8 parts of oxygen united to 1 part of hydrogen. It expresses the same result to say, that 100 weights of water contain 88·89 oxygen plus 11·11 hydrogen. In both cases the elements are in the *same proportion*. Referring the analyses of different bodies to one and the same weight or 100 parts, enables us to compare such analyses, and to ascertain in what respect the quantities of the respective constituent elements vary. This comparison has been made for a vast number of compounds, and has resulted in the knowledge of the "law of proportions" which governs the union of the elements. Of the sixty or more elements with which we are acquainted, carbon, hydrogen, nitrogen and oxygen alone concern us in the vinegar process. Let us proceed to illustrate the above law by their aid.

The law of proportion teaches us that, in every chemical compound the elements unite by weight, each according to a fixed number or its *multiple*.

For carbon the proportional number (called also its "*equivalent*") is 6, for hydrogen it is 1, for nitrogen 14, and for oxygen 8.

It follows then from this law, that if a body consist of carbon, hydrogen and oxygen, the relative weight of the respective elements must be in one of the following proportions:

CARBON.	HYDROGEN.	OXYGEN.
6	1	8
twice 6	1	8
6	twice 1	8
6	1	twice 8
twice 6	twice 1	8
twice 6	1	twice 8
6	twice 1	twice 8
&c.,	&c.,	&c.

We can imagine by this consideration how vast may be the number of chemical compounds containing only three elements in different proportions. The following examples will illustrate the law of equivalents.

Water contains hydrogen and oxygen in the proportion of 1 of the former to 8 of the latter element. Chemists are acquainted with another body containing the same elements but in different proportions. It is called deutoxide of hydrogen, and consists of 1 part of hydrogen in union with 16 (i. e., twice 8) parts of oxygen. Water and deutoxide of hydrogen have entirely different properties.

Again, in carbonic acid the carbon is to the

oxygen as 6 to 16 (i. e., twice 8). Pass carbonic acid over charcoal at a red heat, and a different gas, called carbonic oxide is thus formed. In it the carbon bears to the oxygen the proportion of 6 to 8.

Finally, the table of the compounds of nitrogen with oxygen is illustrative of the law of proportion. It is as follows:—

	NITROGEN.	OXYGEN.
Laughing Gas,	14 +	8
Deutoxide of Nitrogen,	14 +	$2 \times 8 = 16$
Hyponitrous Acid,	14 +	$3 \times 8 = 24$
Nitrous Acid,	14 +	$4 \times 8 = 32$
Nitric Acid,	14 +	$5 \times 8 = 40$

Let us trace now the law of proportion in three "organic" compounds immediately concerned in the vinegar manufacture.

	CARBON.	HYDROGEN.	OXYGEN.
180 parts of Sugar =	6×12	$+ 1 \times 12$	$+ 8 \times 12$
That is	72	12	96
46 parts of Alcohol =	6×4	1×6	8×2
That is	24	6	16
60 parts of Acetic acid =	6×4	1×4	8×4
That is	24	4	32

Observe how sugar, alcohol and vinegar differ; they each contain the same elements, (carbon, hydrogen and oxygen,) but in different proportions. In respect to the carbon the weights are multiples of 6. In sugar the carbon is 12 times 6, in alcohol it is 4 times 6, and in acetic acid or vinegar it is also 4 times 6. The multiples of the hydrogen (1) for the three bodies are 12, 6 and 4, and for the oxygen (8) they are 12, 2 and 4, as may be seen in the table.

It would be well for the reader, if he have not already that knowledge, to make himself acquainted with the chemical "symbols" of carbon, hydrogen and oxygen, and as well to remember how they are used to express the composition of sugar, alcohol and acetic acid. This knowledge will enable me to explain the conversion of sugar to alcohol, and the latter to vinegar in a very simple manner, and so that it may be well impressed upon the mind. This advantage may be gained without the slighest difficulty by resorting to an artificial aid to the memory. The capital letter of the name of each of the elements in question is its "*symbol;*" that is, stands for it, thus C for carbon, H for hydrogen and O for oxygen. Moreover, each of these letters stands for the proportional number or equivalent of its respective element. Thus C stands for 6 equivalent weights of carbon, H for 1 of hydrogen, and O for 8 of oxygen.

These numbers may be very readily remembered. Consider that hydrogen is the *lightest* known substance, it was once universally employed for inflating balloons; it has in fact the *smallest* proportional number of all the elements, it leads the column of equivalents, holds the first rank, is 1; also note that when we write H the pen makes down strokes very like a couple of ones. When we write C for carbon, we make almost exactly a figure 6 its equivalent. Finally the figure 8 is nothing more than two 0's, one above the other. When, therefore, you write O for oxygen, imagine another O on top of the symbol and you have 8, the equivalent of oxygen.

The multiples of the proportional numbers are expressed by a little figure on the right and below the symbol; thus C_4 denotes 4 equivalents of carbon, i. e., since the equivalent of carbon is 6, the above symbol stands for 4 times 6 or 24 weights of carbon. CO_2 is the symbol of carbonic acid. It teaches us that this gas is composed of 6 weights of carbon united to twice 8 weights, that is 16 weights of oxygen. HO is water, 1 hydrogen + 8 oxygen. On the same principle sugar is symbolized thus $C_{12}H_{12}O_{12}$, 12 times 6 weights of carbon, 12 times 1 of hydrogen, and 12 times 8 of oxygen.

The symbol of alcohol is $C_4H_6O_2$. That of acetic acid is $C_4H_4O_4$.

These symbols may be readily remembered. Sugar has a dozen of each of the equivalents of its elements. Acetic acid is ⅓ of the sugar symbol, that is 4—4—4.

Alcohol and acetic acid have the same number of carbon equivalents. We shall find hereafter that alcohol does not lose any *carbon* in passing to vinegar, but is deprived of hydrogen. Almost every one knows that the formation of vinegar is an *oxidyzing* process; the H_6O_2 of alcohol become H_4O_4. In passing into acetic acid alcohol loses 2 equivalents of hydrogen, and takes up 2 of oxygen.

Acohol $= C_4H_6O_2$ becomes acetic acid $= C_4H_4O_4$.

For this ingenious method of illustrating the constitution of chemical compounds by symbols, we are indebted to the celebrated Berzelius.

If the reader be not a chemist he will, doubtless, be much struck with this strange relation existing between numbers and the constituent elements of a chemical compound. If he desires to inquire into the reason of this relation, the atomic theory will afford him all the satisfaction which in the present state of science it is possible to have.

It will be proper here to note well that it is the *proportional numerical relation* of the elements that has been proved by analysis, and not the identical numbers themselves. We have taken 1 to denote the equivalent of hydrogen because it is the smallest of all equivalents, in which case the equivalents of carbon and oxygen are 6 and 8. We might have assigned any other number for H in which case we would find that

6 and 8 standing for carbon and hydrogen would have to be altered proportionally.

In fact the French chemists, because oxygen enters into so many compounds, make it the base of the table of equivalents by calling it 100, in which case carbon becomes 75 and hydrogen $12\frac{1}{2}$. But note that in H : C : O : : $12\frac{1}{2}$: 75 : 100 the numbers are in the same proportion as 1 : 6 : 8.

The Atomic Theory.—Let us imagine a drop of water divided into smaller ones, and each of these into still smaller ones and so on. It will not be long before we reach a drop of the smallest size appreciable by our senses; but we may go on dividing in imagination and when the mind, bewildered at last, pauses, let us ask the question "Can we go on dividing *forever;* or must we at length reach a point at which the drop is *no longer susceptible of division?"* The atomic theory supposes the latter alternative, viz. : that bodies cannot be infinitely divided. A particle incapable of further division is at length reached to which is given the name atom, the etymology of which is *indivisable.*

In this example we reach the *atom* of water which we say is indivisable, that is *as water;* we may indeed separate it into hydrogen and oxygen, but it is then no longer water. Hydrogen, oxygen and all the elements have their respective atoms, which cannot of course be divided; if they could the element would be no longer an element but a chemical compound.

Now these atoms *must have weight*, for dividing a body does not destroy or diminish the aggregate weight of the portions into which it is divided. The question very naturally presents itself whether the atom, say of hydrogen, has the *same weight* as the atom of oxygen, or as the atom of any other element? In other words Have the atoms of the different elements the same weight? We cannot now, and probably never may tell the *actual weight of any atom*, for such weight is far too small for the sensibility of our most delicate balances. We may, however, by the aid of the atomic theory form a very reasonable supposition as to the relative weight of the atoms. We have every reason to believe that whatever an atom of hydrogen may weigh, an atom of carbon weighs 6 times as much, and an atom of oxygen 8 times as much. The atomic theory *assumes* that in chemical combination it is the *atoms* that unite; that one or more atoms of one element unite with one or more atoms of another element. Furthermore, that the equivalent numbers represent the *relative weights* of the atoms. For example, one atom of water contains an atom of hydrogen, and an atom of oxygen, and the atom of oxygen weighs 8 times as much as the hydrogen atom. If we analyze say 9 grains of water, we will obtain 1 grain of hydrogen and 8 grains of oxygen. Now I would like the reader here to note the uncertainty of the *numerical result* of the theory. The *analysis* is perfectly correct; one-ninth weight of any given quantity of water is hydrogen, the rest is oxygen. In the analysis we weigh millions of atoms. What right have we to say that 1 atom of water contains 1 atom of hydrogen and 1 atom of oxygen? We might say that it contains 2 atoms of hydrogen and 1 atom of oxygen, provided we assign $\frac{1}{2}$ as the weight of a single atom of hydrogen. The atom of oxygen would then weigh 16 times as much as the hydrogen atom. In fact some chemists have made this supposition. To say which supposition agrees best with the present state of chemistry, involves a consideration of its whole domain.

Whatever we may assign as the difference between the weight of an atom of hydrogen and one of oxygen, there is no question as to the numbers resulting from the analysis of water, nor is there anything unreasonable in the supposition that in a chemical compound the *atoms* unite, which atoms have a different weight for every element.

Without assigning the grounds for our supposition, let us admit with the majority of chemists, that water is composed of equal atoms of H and O, that the oxygen atom weighs 8 times as much as the hydrogen atom, and that

CHEMICAL PRINCIPLES. 45

the nitrogen atom weighs 14 times as much. It will then follow as a necessary result from the atomic theory that, the elements must unite in the proportion of their equivalent numbers or of multiples of those numbers.

If the atom of O weigh 8 times as much as the hydrogen atom, and if water contains *equal atoms* of hydrogen and oxygen; 9 pounds, ounces or grains of water must contain 1 pound, ounce or grain of hydrogen and 8 pounds ounces or grains of oxygen. Deutoxide of hydrogen contains the same elements as water. We may expect to find one of the elements as a multiple of its equivalent; this is the case. The union of 1 pound of hydrogen with 16 pounds of oxygen gives rise to 17 pounds of deutoxide of hydrogen; thus by the atomic theory

(H) 1 atom of hydrogen weighs, . . . 1
(O_2) 2 *atoms of oxygen weigh twice* 8 = . . 16
(HO_2) 1 ATOM OF DEUTOXIDE OF HYDROGEN weighs, . 17

The difference therefore, between water and deutoxide of hydrogen is that water contains 1 atom of oxygen united with 1 atom of hydrogen, while in deutoxide of hydrogen 2 atoms of oxygen are joined to one of hydrogen. The following according to the Atomic Theory is the

TABLE OF NITROGEN WITH OXYGEN.

	In equivalents.
	N O
(NO) Protoxide of Nitrogen,	
1 atom N + 1 atom O	14 + 8
(NO_2) Deutoxide of Nitrogen,	
1 atom N + 2 atoms O	$14 + 2 \times 8 = 16$
(NO_3) Hyponitrous Acid,	
1 atom N + 3 atoms O	$14 + 3 \times 8 = 24$
(NO_4) Nitrous Acid,	
1 atom N + 4 atoms O	$14 + 4 \times 8 = 32$
(NO_5) Nitric Acid,	
1 atom N + 5 atoms O	$14 + 5 \times 8 = 40$

I have thus purposely described the Atomic Theory by way of *speculation;* for it is a speculation, and however reasonable and explanatory in a simple manner of chemical phenomena, is not in the present state of the science, *susceptible of proof.*

Let us in conclusion, take a practical example of this theory applied to alcohol, which is expressed by the symbol $C_4H_6O_2$. In this body, by the theory, four atoms of carbon, five of hydrogen, and two of oxygen are conjoined, with the result of one atom of alcohol. By the atomic theory, we can infer the analysis: for since the weights of the atoms of CHO, are respectively 6 : 1 : 8; $4 \times 6 = 24$ pounds of carbon joined to $6 \times 1 = 6$ pounds of hydrogen, and $2 \times 8 = 16$ pounds of oxygen give 46 pounds of alcohol.

The compounds belonging to the department of organic chemistry, are signalized by the readiness with which they fall apart under the influence of reagents, to form new combinations. This behavior may be expected from the atomic theory. Such compounds are like the pile in the child's house of cards. With skill we may remove a card, (atom,) or two, forming a structure of different shape; but there are cards, which if removed, will involve a complete ruin of the edifice. Besides, if two such houses are in contact, the fall of one will result in the fall or change of its neighbor, as in the chemical case of fermentation, explained by Liebig's theory.

CHAPTER II.

SUGAR.

ALL the vinegar used in the household arises from a transformation of alcohol, which itself results from a chemical change experienced by sugar. In the majority of instances, vinegar manufacturers by the quick process employ ready formed alcohol, which they buy as alcohol, spirits, high wines, whisky, or some similarly distilled liquid. It might, at first sight, appear useless in this work, to devote a chapter to sugar, and part of another to the manufacture of a fermented liquid. It has, however, been deemed advisable to treat these subjects in a general manner, not only that a few may avail themselves of what is said of them, but that a clearer idea may be presented of the relation of sugar and alcohol to vinegar; for there are some who seek to improve their manufactured vinegar by the addition of a saccharine substance, not aware that the alcoholic and the acetic fermentations are entirely different processes, and to be effected, require different conditions, whether of temperature, nature of the ferment, &c.

Although, according to the plan of this work, Part I. is devoted to theoretical considerations;

in the following chapters upon sugar and alcohol a few practical operations will be discussed. In the present chapter, we will, after considering the chemical nature and properties of the sugars and allied bodies, learn how they may be obtained in a liquid capable of fermentation. In the chapter on alcohol we will proceed with the change of sugar to alcohol, both theoretically and practically, besides giving what alcoholic information the vinegar maker ought to have, if he employ in his process spirits already manufactured.

The following table contains the principal sugars and allied bodies, that is, bodies from which sugar may be made, together with their chemical composition:

THE SACCHAROID BODIES.

Cellulose { The cellular tissue of plants, woody fibre, } $C_{12}H_{10}O_{10}$.
Starch . . . . $C_{12}H_{10}O_{10}$.
Gums { Arabic, ... $C_{12}H_{10}O_{10}$.
Dextrine, ... $C_{12}H_{10}O_{10}$.

Sugars
{ Cane sugar crystalized, . } $C_{12}H_{11}O_{11}$.
{ Supposed to be arranged, } $C_{12}H_9O_9 + 2HO$.
Fruit sugar, (uncrystallzable,) . $C_{12}H_{12}O_{12}$.
{ Glucose or rasin sugar, called improperly grape sugar, } $C_{12}H_{14}O_{14}$.
Milk sugar crystalized
  arranged as, { $C_{12}H_{12}O_{12}$.
  $C_{12}H_{10}O_{10} + 2HO$.
  or as some suppose, { $C_{24}H_{24}O_{24}$.
  arranged as, $C_{24}H_{19}O_{19} + 5HO$.

The formula in the last column denotes the number of atoms or equivalents which each body contains; thus in cellulose, $C_{12}H_{10}O_{10}$, there are 12 equivalents or atoms of carbon united with 10 of hydrogen, and the same number of oxygen. It will no doubt surprise the non-chemical reader to find bodies as unlike in their general properties as wood, starch, gum, and sugar classed together, but let him note how analogous is their *atomic* constitution. These bodies all contain 12 atoms of carbon, and their hydrogen and oxygen atoms are in equal proportion; that is, in the proportions proper to form water. One would suppose that we might transform one member of the table into another by the mere addition or abstraction of one or several atoms of water. This is generally the case; indeed, we may easily remember them all, if we say that the saccharoid substances contain 12 atoms of carbon, and respectively 10, 11, 12, and 14 atoms of water.

These bodies, with the exception of milk sugar, all by fermentation fall apart into exactly the same substances, namely: alcohol, carbonic acid, and water; the sugar *directly*; the cellulose, starch and gum, after having first experienced, by a simple chemical process, a change into sugar. It will perhaps be asked, how cellulose, starch, and gum can have the same atomic constitution, and yet be different bodies? The answer is ready. The given formulæ denote merely the *number* of *atoms* of the constituent elements, not the *arrangement* or grouping of the atoms, which is doubtless, different for each body. For example, the formula for cane sugar is $C_{12}H_{11}O_{11}$; but we have chemical reasons for supposing that these

CELLULOSE. 51

atoms are grouped thus: $C_{12}H_9O_9+2HO$. Though it may be said, for the purpose of remembering the formulæ of the table that *all* of their H and O are contained as water; this statement is, strictly speaking, incorrect.

It should be stated, that while the *proportion* between the atoms of CH and O in the foregoing table is firmly established, there is a difference of opinion as to the *actual number* of atoms of each element present. Thus cellulose might be $C_{24}H_{20}O_{20}$; $C_{36}H_{30}O_{30}$; $C_6H_5O_5$, &c.; for in all these formulæ, CH and O are in the same proportion, namely as 12:10:10. In fact, some chemists regard milk sugar to be $C_{12}H_{12}O_{12}$; others as double this formula, $C_{24}H_{24}O_{24}$. The formulæ given in the table are those adopted by the majority of chemists. To proceed with the subject—cellulose, starch and gum, however their atoms may be grouped, have the same general atomic constitution $C_{12}H_{10}O_{10}$.

CELLULOSE.

This body is the material of the myriads of cells which make up every vegetable body. It is sometimes, though improperly, called lignine. The cells themselves are filled in some instances with granules of starch, as in the potato and certain roots and seeds. In other cases their walls are encrusted with a firm woody matter, the *true lignine*. Wood owes its stiffness to this lignine, of which the composition is unknown. In young plants the cells contain a fluid or viscous matter holding in solution mineral salts, gums, gelatinous, and albuminous matter, &c., &c. In oleaginous seeds, as of the olive, flax, various nuts, &c., the cells contain besides the above mentioned substances, others of an oily nature. These cells may be seen in every vegetable body by examining a thin slice of it with the microscope; in the pith of the elder they may be readily perceived with the naked eye. In the asparagus they are also thus visible, and present the appearance of lengthened cylinders. Vegetable cells present a varied form. In hemp and flax they are long and cylindrical; in the cotton fibre they are flattened, and like a twisted ribbon. Cotton and linen afford us the purest cellulose, especially when in the form of paper, and old linen or cotton rags, because the chemical and mechanical processes which they have been subjected to in their manufacture and "wear" have removed the foreign matters which are more destructible than cellulose. Pure cellulose is white, translucent, insoluble in water, alcohol, ether, or the oils. With concentrated nitric acid, it experiences a change, giving rise to an eminently explosive substance—gun cotton. Its behavior with oil of vitriol concerns more immediately the vinegar manufacturer. This acid changes it to a *gum* called dextrine, which, as we shall see, by boiling with a weak solution of the same acid, passes into glucose or raisin sugar. Any one may perform this interesting experiment as follows:

Add in a wedgewood mortar, oil of vitriol to

half its weight of dry linen, cotton rags, or paper. The mixture must be made very gradually, or the heat evolved will char the mass. After all the acid has been added, rub up the pulpy mass with the pestle, and let it stand for several hours; after which, add *very gradually*, its bulk or more of water; stir up well, and warm for a short time; then filter off the solution. The woody fibre has thus been converted into " *dextrine;*" but the acid in the solution must be got rid of. This is effected by adding thin lime whitewash; but not enough to completely neutralize the acid. The neutralization is completed by means of chalk or powdered limestone. If, after once more filtering, we evaporate the solution, the dextrine remains as a gummy mass. If we desire sugar instead of dextrine, we must before neutralizing the acid, boil the solution for three or four hours, replacing the water as it evaporates. The dextrine passes probably first into fruit sugar; but certainly at last into raisin sugar. The process is completed as before by neutralizing the acid, filtering and evaporating to the crystalizing point. Linen rags may thus yield more than their own weight of sugar; but the process is comparatively an expensive one. As the sugar from wood is fermentable, and the resulting alcohol susceptible of the vinegar process, we need not be surprised to find that vinegar is one of the products of the dry distillation of wood. But the manufacture of vinegar from wood by destructive distillation is many times cheaper than the chemical process.

STARCH.

It has been said that some plants contain in certain of their cells, starch. If with a sharp knife we take the thinnest possible slice of a potato, so thin that, it is transparent on the edges, and then placing it upon the point of a needle, stir it gently in water, so as to wash off the adherent matter, the microscope will afford an ocular demonstration of the statement. Place the slice upon a slip of glass with a drop of clear water upon it, and covering it with a piece of thin glass, such as is used in microscopy, examine its magnified image. A most beautiful and interesting picture will present itself. The cellular structure of the vegetable is manifest; some of the cells are empty, their contents have been discharged by the cutting and subsequent washing. Others are filled with transparent, shining, egg-shaped particles of various sizes; they are the *starch "globules."* Some of them are exactly egg-shaped; others are more or less irregular. A close inspection will manifest upon each a *"hilum,"** around which the starchy matter appears to have been deposited in eccentric layers. If the glasses between which the starch globules

---

* Hilum, a mark. In botany the mark on a fruit or seed indicating where the stem was attached.

lie be pressed together to break some of the starch globules, the fracture passes through the hilum as if the globules were weaker at that point. If we heat *very gradually* the.glass on the side of the starch 'deposit, the peculiar action of warm water upon starch will be exhibited. The starchy layers in each globule will begin to exfoliate about the hilum, demonstrating that the matter is in layers. By greater heat the globules disappear, having become completely exfoliated and transparent. By close attention we can perceive the now shrunken membranous sac which originally enveloped the starch globules.

Viewed by polarized light, starch is a beautiful microscopic object.* In a certain position of the polarizing apparatus, every globule has a black cross depicted upon its shining white surface. The intersection of the cross coincides with the hilum. Finally, if we place the starch of different plants under the microscope, we will find that although they bear the general characteristics which I have described in potato starch, they are so different in average size and shape that we may assign the starch to its plant, or rather plant family, by such microscopic examination.

Starch enters largely into the food of man, and of every graminivorous animal; accordingly we find it in some portion of almost every plant. It

* Modern improvement has so cheapened the microscope as to place this interesting instrument within the reach of almost every lover of nature. See every Optician's Catalogue.

is eaten, not only associated with gluten, in the different varieties of bread or cakes and in certain vegetables, but is also prepared separately for purposes of food, as in sago, tapioca, arrowroot, farina, corn starch, &c.

Corn starch is from maize; farina is the starch of wheat; sago that of the pith of the trunk of a species of palm tree; tapioca and arrowroot are starch obtained from two different kind of roots growing in the West Indies.

Thus this important substance occurs in the root, stem, seed, or fruit of a great many plants. In the root, as of potato, arrowroot, and the plant yielding tapioca; in the stem of the sago palm; in the seed, as of rice, peas, beans, wheat, barley, rye, oats, and the cereals generally; in the fruit, as of banana, plaintain, bread fruit, &c. The amount of starch contained in these different substances varies according to the kind and culture of the plant. Thus potatoes contain from 14 to 20 per cent. of starch. The following is the composition of an average variety:

Starch,	20·0
Cellulose,	1·7
Gluten,	1·5
Gum, sugar, and oil,	1·3
Mineral salts,	1·0
Water,	74·5
	100·0

Wheat contains a percentage of starch of from

55 to 77, and from 7 to 20 gluten; the remaining constituents being gum, oil, cellulose, mineral salts, and water.

Rice contains 86 per cent. of starch.

Preparation of Starch.—In preparing starch from these different plants, the simple method always obtains of rupturing the cells, washing out the starch globules thus set free, and collecting them after they have settled in the water, than which they are heavier, having a specific gravity of 1·53.

In obtaining starch from wheat and similar seeds, the gluten presents an obstacle to be overcome. Several methods are adopted for liberating the starch globules from their cells. Sometimes the wheat is suffered to swell in water until soft, and then placed in warmer water to ferment, when the starch may be squeezed out by pressure in bags, or by grinding under vertical edge-stones. Some prefer to crush the seed between iron rollers before steeping. In all cases the starch is liberated, and much of the gluten got rid of by fermentation. The globules are then suspended in water, passed through fine sieves, and permitted to settle repeatedly, and at the last suffered to remain in the water from four to five days. In settling, the round form of the starch globule is of advantage, for it enables it to reach the bottom of the vat much sooner than the light, flattened shreds of cellulose. The deposits of gluten, cellulose, &c., resting upon the stratum of starch, are called slimes. They are rejected and serve for food for animals.

The starch is at last removed from the water and placed in perforated canvas lined boxes, where its water for the most part drains off. The mass is then broken up into rectangular plates, which are dried, first upon half burned bricks, and finally in stoves. During dissication the plates split up into the prismatic columnar form in which we find the starch of commerce. Wheat yields an average of from 35 to 40 per cent. of good starch.

Potato starch is sometimes used by the vinegar maker. This he prepares himself, by reducing the vegetable to a pulp in a rasping machine, under a moderate stream of water. The pulp is squeezed and stirred by the hand, or by machinery, upon a fine wire or hair sieve, under thin jets of water. The pulp remaining is re-rasped. The milky water holding the starch in suspension, is collected in a vat, where it remains for a time. The supernatent water is brought into a second and thence into a third vat. The starch settles in these different vats of different fineness and purity. It contains about 33 per cent. of water, and is called green fecula. It may, of course, be treated like wheat starch and obtained dry; but the vinegar maker employs it in the moist condition, preserving it under a layer of water, which is renewed from time to time, to prevent fermentation and at length converting it into raisin sugar and alcohol by processes to be described hereafter. The product of starch from the potato, is from 17 to 18 per cent. of the weighed tuber.

Chemical nature of starch.—Chemists are acquainted with at least three varieties of starch. 1. Starch proper, such as has been just described, existing in the potato, wheat, &c., and which is always intended when speaking of starch in this work. 2. Inuline, found in the roots of several plants, as elecampane, the dahlia, &c. 3. Lichen starch, obtained from several lichens and algæ, among which may be mentioned Iceland moss and carrageen or Irish moss.

These three varieties agree in their insolubility in cold water, alcohol, ether or oil, and in being, by the aid of dilute sulphuric acid or diastase, converted into sugar.

They differ in their behavior with *hot* water and iodine; thus: 1. Ordinary starch gives with hot water a mucilaginous liquid, jellying when cold, and colored intensely blue with iodine. 2. Inuline gives a granular precipitate when its boiling solution is cooled, and which is colored *yellow* by iodine. 3. Lichen starch yields a gelatinous mass when cold, which is colored brownish gray by iodine.

To return to ordinary starch; if we place some in cold water, no change takes place, but upon boiling the starch, layers exfoliate, having burst the membranous sac enveloping the starch globule. If too much water has not been taken, a gelatinious mass, the starch of the laundress, remains on cooling. It has been supposed that an actual solution of the starch does not here take place, because if the apparent solution be frozen the starch all separates.

A watery solution of iodine colors intensely blue, not only the starch globules and starch paste but also the water in which starch has been boiled. *Cellulose* is not colored blue by iodine, but acquires this property by a short contact with oil of vitriol, which seems to prove that woody fibre in its passage to dextrine and sugar, assumes something of the starchy nature.

Starch itself may be transformed to *gum dextrine* by heat, acids, and by *diastase*. The gum formed by heating dry starch can better be treated under the head of dextrine. As a prolonged action of acids and diastase, result in the change of dextrine to raisin sugar, I shall defer the consideration of these phenomena to the portions of this chapter which treats of sugar.

I have used several times the word diastase; it may be well to conclude the subject of starch by a description of this wonderful body.

Every living seed when placed under the conditions of moisture and warmth favorable to its development, sprouts, during which process "*diastase*" is formed, probably from the albuminous matter of the seed. This diastase has the remarkable property of converting starch first into a gum called dextrine and then into sugar. In brewing, this property is utilized. The operation of malting brings about an artificial germination of barley or rye, whereby diastase is formed, which acts upon the starch of the seed, converting it into the fermentable bodies, gum and sugar. To obtain and test the properties of diastase, we have only to add alcohol to a filtered watery solution of crushed malt, which precipitates diastase along with other matter. If we place this sediment in contact with starch and water at a temperature of from 140-149° Fah., the transformation of starch to gum first and then to sugar may be observed.

Diastase thus plays an important part in the economy of nature. A seed is like the fecundated egg of a fowl which contains within its shell everything needed for the formation of a perfect bird, which under favorable circumstances breaks from its imprisonment and leads a life of an entirely different character from its embryonic condition. The plant roots itself immovably in the earth, and its food comes to it from earth and air through its rootlets and leaves. Its food, besides mineral salts, is carbonic acid and water, which it decomposes *in the light*, appropriating the carbon, hydrogen and some oxygen, to form compounds necessary to its existence, and breathing out the greater part of the oxygen. The plant is an organism essentially opposite to the animal; in fact, they are complementary to each other, for each yields what the other requires for its life, and together they complete the grand circle of vitality. The animal nourishes itself with the hydro-carbon matter of the plant, oxidizing these in its body and breathing them out in the form of carbonic acid and water necessary to plant life; while, on the other hand, the animal could not live without oxygen, the breath of the plant. The plant to develop itself by the aid of carbonic acid and water, needs to root itself in the ground and rear its stem in the air. How then can the seed buried in the dark earth gain this advantage? It contains in its envelope enough matter to shoot a rootlet downward and to elevate a delicate stem and leaf to the air, and diastase, formed during the germination, appears to play the part in it of converting its starch first into soluble gum and sugar, from which the insoluble cellulose of the cells is formed, upon the walls of which the rigid lignine is deposited, giving rise to the stiff wood thrusting itself downward in the radicle and upward in the stem. Starch dextrine and cellulose all contain $C_{12}H_{11}O_{11}$. The life process I am describing, arranges by the aid of diastase, these atoms in the requisite manner to effect the change from starch to wood. With the first root and leaf the plant is born, it shifts for itself, takes in matter from quarters exterior to itself, and advances to the perfection of its development, transmitting its life force to a subsequent generation, by bearing seed after its kind.

THE GUMS

constitute a class of bodies, distinguished in the popular mind for the adherent, sticky, or so called gummy nature of their solutions, and are employed in the arts for pasting, thickening colors, as mordants in calico printing, and for giving gloss and finish to certain woven fabrics. Gum arabic and tragacanth, as well as the *cerasine* which exudes from the trunk of the cherry, peach, plum and similar trees, are well known. Dextrine the artificial gum, prepared from starch, takes rank with these.

The gums have the general atomic constitution $C_{12}H_{10}O_{10}$, that is as far as the number of atoms of their elements go, they are identical in composition with each other, with woody fibre and with starch. They all differ in certain chemical and physical properties, which is attributable to the arrangement of their atoms, and which difference does not concern our present purpose. It suffices to know their points of resemblance, viz., that none of them crystalize; that all when heated with a dilute solution of mineral acid, (sulphuric,) are transformed into sugar, which may be converted into alcohol and then vinegar; that when pure they are not colored by iodine; and that, although more or less soluble in water, they are insoluble in strong alcohol. The latter property enables us to separate sugars and gums, for if we add enough strong alcohol to the mixed solution, the sugar will remain in solution, and the gum will assume the form of a sediment.

DEXTRINE

arises from the transformation of starch, in three ways; by acids, by the action of diastase, and by heat. In its adhesive property and general appearance when pure and dry, it much resembles gum arabic. As it costs much less than gum arabic, it is extensively employed in the arts as a substitute. It has received its name, which means "right-handed," from its action of twisting the plane of polarized light to the right hand to a greater degree than any known body. I will endeavor at the close of the chapter, to give a popular explanation of this effect on polarized light.

Dextrine may be prepared by the following different methods:

1st. By boiling starch with almost any dilute acid. A little sulphuric acid added to a boiling gelatinous starch paste will, in a very short time, make a solution as limpid as water. The change to dextrine has been thus effected. We can readily ascertain when the starch is all gone, by adding every few minutes to a small portion of the liquid, a few drops of iodine solution, which strikes a blue color with starch, but does not affect the dextrine. It is important to stop the boiling at the precise point when iodine no longer colors the solution, since prolonged boiling changes the dextrine to raisin sugar. Thus, if 15 parts of starch, 60 of water and 6 of oil of vitriol, be boiled together for four hours, replacing the water as it evaporates, the dextrine at first formed is entirely converted into sugar. The process can be interrupted at any moment; the acid neutralized by chalk, and the dextrine, sugar, or their mixture, recovered from the filtered solution by boiling it down to a syrup.

2d. *Preparation by Diastase.*—It has been observed that diastase effects the above change of starch to dextrine and sugar. The following simple experiment may be performed by any one: Powder a small quantity of malt (germinated barley), and soak it for several hours in a little water of the temperature of 80–85° Fah., squeeze the water from the parts through linen and filter; the result is a solution of diastase. One part by weight of dry diastase is capable of transforming 2000 parts of starch to dextrine, and subsequently to sugar. Hence, one part of diastase is as effective as 30 parts of sulphuric acid. If the diastase solution be added to gelatinous starch kept at a temperature between 150–170° Fah., the transformation to dextrine will speedily take place, a delay of several hours at this temperature will complete the change of dextrine to sugar, and we can, as before, ascertain when all the starch is gone by the use of iodine solution. The action of the diastase may be arrested at any point, by raising the temperature to the boiling point.

3d. *Preparation by Heat.*—Dry starch raised to the temperature of 400° Fah., is converted into dextrine. Except that it is of yellowish tinge its appearance is unchanged; it dissolves readily in water yielding a gummy solution. Dextrine used in the arts for gum purposes is thus prepared; it is called British gum. A *trace* of acid added to starch enables British gum to be made at a much lower temperature than 400°. For example: if 1000 pounds of starch be moistened with a mixture of 300 pounds of water and 2 pounds nitric acid, and after spontaneously drying be exposed for one or two hours in stoves to a temperature of 212–230° Fah., the transformation will be complete, and all the acid will have evaporated.

British gum is not, like dextrine prepared by sulphuric acid or diastase, transformable into sugar.

THE SUGARS.

The sugars constitute a class of bodies most important for man's food. Sugar is not only in the earliest nourishment which the mammal requires; not only is it eagerly sought by the young, and enjoyed by men of all ages in the various fruits and vegetables constituting their food, but enormous quantities of pure sugar are manufactured and consumed every year, as may be seen by Dr. Stolle's table quoted in the "Chemistry of Common Life."

THE SUGARS. 67

	Millions of pounds annually.	Per centage of the whole Sugar production.
Cane Sugar,	4527	87.7
Beet "	362	7.3
Palm "	220	4.2
Maple "	45	0.8
		100.0

In the following countries the assigned quantities of sugar are used yearly in proportion to their respective populations:

Countries.	Pounds per head annually.
Russia,	1½
Belgium,	5
France,	7½
Great Britain,	28
Venezuela,	180

The amount to each inhabitant of the United States is not stated, but it is safe to say that the proportion far exceeds that of Great Britain.

The sugars cited in the first table are identical in nature with cane sugar, although obtained from such different plants. If asked for a simple definition of the true sugars I would say that they are bodies of sweet taste, soluble in water, less soluble in alcohol, yielding alcohol and carbonic acid by the vinous fermentation, and having in every atom of sugar twelve (or a multiple) atoms of carbon, and hydrogen and oxygen in equal atoms, that is in the proportion proper to form water. This last characteristic is expressed by giving sugars the name "*carbohydrates*" i. e. "*charcoal waters.*" My definition excludes mineral sweets, as sugar of lead and hyposulphate of silver. It also shuts out certain vegetable sweets as the manioc, and liquorice which are not "*carbohydrates*," nor yielding alcohol and carbonic acid by yeast fermentation.

All sugars are not identical with cane sugar. Here are the different varities:

Cane Sugar,	. . . . .	$C_{12}H_{11}O_{11}$
Fruit "	. . . . .	$C_{12}H_{12}O_{12}$
Raisin "	. . . . .	$C_{12}H_{14}O_{14}$
Milk " $\Big\}$ either $\Big\{$	. . . .	$\Big\{ C_{12}H_{12}O_{12} \Big\}$
or	. . . .	$\Big\{ C_{24}H_{24}O_{24} \Big\}$

It is unfortunate that the sugars form so few definite crystalizable compounds with other bodies. For this reason, analysis cannot show whether the number of atoms assigned actually belongs to the sugars, although its results are certain with respect to the *proportion* existing between the numbers of atoms of carbon, hydrogen and oxygen.

Milk Sugar.—Let us consider this variety at once so as to have done with it. It enters in no respect into the vinegar manufacture and finds a place here solely because the Tartars ferment the milk of their mares to form the alcoholic drink Kouhmiss, which when distilled yields the spirit called arrack. This sugar is contained in the milk of all mammalian animals, and has been detected, it is said, although the fact has been denied by Dessaignes, in the acorn.

In Switzerland and in other cheese-making

countries, it is manufactured for sale to a small extent, by evaporating to crystalization the whey after separating the curd for cheese. The homœopathist employs it to infinitely dilute his drugs, by triturating a small portion of them with a large quantity of milk sugar.

This body crystalizes in white four sided prisms, is gritty to the teeth, less sweet and less soluble than cane sugar, requiring 4 parts of hot and 7 of cold water. Dilute acids transform it to raisin sugar. It does not ferment with yeast alone, but dissolved in the milk with butter and curd readily yields carbonic acid and alcohol. A characteristic product of its fermentation is *lactic acid*. Chemists are not agreed whether it contains 12 or 24 of the atoms respectively of its elements.

Cane, Fruit and Raisin Sugars are so well known by experience if not by name, that it is not too much to say that every civilized individual is in the habitual use of all of them. All the sugar sold as such in the *solid* state is of the same variety as cane sugar. Fruit sugar is enjoyed in molasses and in acid fruits, and raisin sugar in raisins and other acid fruits which have been dried.

1st. *Cane Sugar* is found in the sugar cane, sorghum, stalk of maize, the wounded top shoot of the palm, American aloe, &c., in the ascending sap of the maple, in beets, melons, cocoa, pine apples, bananas, the nectaries of flowers, and in the juices of parts of many other vegetable productions. It is a singular fact that those parts of the plant most remote from the leaves contain juice the richer in sugar, which appears to show that the sugar experiences a change in the branches to cellulose or the fibre of wood. We have learned how wood may be converted into sugar; here is perhaps, an example of the transformation of sugar to wood.

2d. *Fruit Sugar* is found in acid fruits, as the grape, gooseberry, strawberry, apple, peach, currant, pear, cherry, &c. Also, in honey, where it results probably from a transformation in the bee of the cane sugar of the nectaries of flowers. It also results from the change of cane sugar by acids or by ferments, and is well known to us in molasses, the larger portion of which is fruit sugar.

3d. *Raisin Sugar* which bears also the name "*glucose*," may be seen in the form of little whitish grains covering the surface of raisins and of prunes. It is generally called "*grape*" *sugar;* a misnomer, for grapes contain chiefly *fruit* sugar, which is transformed into raisin sugar by the process of drying and by time.

Since fruit sugar becomes converted into raisin sugar by standing, we often find liquid honey become by age, a crystaline mass of raisin sugar. Raisin sugar is also the result of the transformation of woody fibre, starch and gum by acids, or in the case of the last two bodies, by diastase.

As might be supposed it frequently occurs with fruit sugar in the juices of acid fruits in honey, &c. It has been noticed in dry seasons as an exudation upon the leaves of some forest trees. In the animal kingdom it has been found in the white of an egg, in the liver and blood, and always in the urine of persons suffering from the disease "diabetes mellitus." In this disease the patient frequently passes over 30 pints of urine per day, containing 3 pounds 4½ ounces of raisin sugar. Milk sugar may be converted by dilute acids into raisin sugar.

Let us now, as briefly as may be, consider these three kinds of sugar.

1st. *Cane Sugar.*—This variety is manufactured largely from the juice of the cane, beet root, and sap of the maple. The average composition of the sugar cane is

Sugar,	18–22
Water and gluten,	71
Woody matter,	10
Salt,	1
	100

That of the European cultivated sugar beet,

Sugar,	10½–14
Water,	81½
Fibre,	5
Gluten, &c.,	3
	100

While from 6 to 7 per cent. of cane sugar are

extracted in Europe from fresh beet root, not more than 6 to 6½ result generally in the West Indies from the sugar cane.* This difference is due to the superior scientific skill applied to the former manufacture.

The loss of sugar is attributable to the change which the juice experiences in its treatment, by which 3 per cent. of sugar are converted into molasses (uncrystalizable fruit sugar), 2½ are lost in the skimmings, and 6 per cent. left in the canes. The molasses of the beet owing to its unpleasant taste is not available directly for sweetening purposes.

Cane sugar crystalizes readily and beautifully, as may be seen in sugar candy. The density of these crystals is 1·6. It dissolves in ⅓ only of its weight of cold water. Eighty parts of boiling absolute alcohol will dissolve one part of sugar, which is almost entirely separated on cooling. Weaker alcohol having a strength of 83 per cent, will dissolve ¼ its weight of sugar at the boiling temperature.

The action of heat on sugar is worthy of notice. Raised to a temperature above 320° Fah., it melts to a viscous fluid, which when suddenly cooled is like glass, brittle, transparent and non-crystaline. Its atomic constitution is as before $C_{12}H_{11}O_{11}$. If this candy be kept for some time in the air it becomes gradually crystaline, to produce which effect its molecules must gradually shift

* In Cuba 10 to 12 per cent.—*Johnston.*

their relative positions in the solid mass. Chemistry furnishes us with several examples of this singular phenomenon in solids. This molecular rearrangement takes place suddenly if candy while yet warm be pulled, the mass becoming white and opaque. The temperature rises during the operation as if a condensation were taking place.

Sugar kept for some time at 356° Fah., is radically modified. It loses the power of crystalizing from its solutions. Between 378–396° it parts with two atoms of water, becoming $C_{12}H_9O_9$, and *caramel*, a black substance, no longer sweet or fermentable, but deliquescent and readily dissolving in water, to which it imparts a deep brown color. Caramel is employed to give vinegar made from alcohol and which is as limpid as water, the fine natural color of wine or cider vinegar. When sugar is kept in contact with almost any acid, even with dilute ones, it becomes uncrystalizable and presents a very great analogy to fruit sugar.

Cane sugar forms several compounds with mineral salts; some of them are very deliquescent and refuse to crystalize with the pure sugar remaining in the molasses. As the juices of sugar bearing plants contain a certain quantity of mineral matter, we may readily perceive how a loss of crystalizable sugar is due to this property.

If we examine molasses, we find that it contains, 1st. Uncrystalizable sugar, arising from the action of heat in the boiling process, and from the action of acids and ferments upon the juice. 2d. Crystalizable sugar kept in solution by its combination with mineral salts. 3d. Free crystalizable sugar which cannot readily separate in crystaline form from the viscous molasses, which impedes that motion of the sugar molecules so necessary to crystalization. This portion of sugar, however, separates from the molasses by long standing. The analysis of crystalized cane sugar, yields the formula, $C_{12}H_{11}O_{11}$, in which the atoms are supposed to be grouped $C_{12}H_9O_9+2HO$, because we may form a compound of oxide of lead and sugar, containing $C_{12}H_9O_9+2$ atoms of oxide of lead; and again we may take this lead compound and substituting therein 2 atoms of water for the two of oxide of lead obtain crystalized sugar having the original formula $C_{12}H_{11}O_{11}$.

Cane sugar is susceptible of the alcoholic fermentation, but it must pass first into fruit sugar. This change may take place by acids, either already present or generated in the fermentable liquid, or by the ferment itself.

The juices of sugar yielding plants contain everything necessary for fermentation. When exposed to the air at a slightly elevated temperature, substances are generated which transform cane to fruit sugar, and a ferment arises which transforms the latter into alcohol.

To affect the alcoholic transformation, pure sugar requires to be diluted with water to a strength of from 10 to 12 per cent. and a ferment then added.

2. *Fruit Sugar.*—This sugar, found in the juices of acid fruits is very sweet and uncrystalizable. It occurs in the ascending sap of the birch, and in the descending sap of the maple. It is frequently associated with cane sugar, as in honey, grapes, &c. It may be readily obtained, by saturating the acid juices with chalk, then boiling with white of egg, which in coagulating separates certain mucilaginous substances. The filtered liquor is then evaporated by a gentle heat. When dry it has the appearance of gum, does not crystalize, and soon becomes liquid by attracting moisture from the air. It dissolves very readily in water, also freely in alcohol of 33 per cent., but is nearly insoluble in absolute alcohol. A ferment added to its watery solution at once induces the alcoholic transformation.

If a syrup of fruit sugar be kept for a long time, a portion is transformed into *raisin* sugar, which crystalizes out in grains. The dried sugar also becomes crystaline by long standing, having first attracted moisture from the air. Chemists explain this change of fruit sugar, ($C_{12}H_{12}O_{12}$,) to raisin sugar, ($C_{12}H_{14}O_{14}$,) by the assumption of 2 atoms of water, 2HO.

When a watery solution of cane sugar is boiled for a considerable time, it refuses to crystalize; the result is supposed to be a mixture of cane sugar with fruit sugar which has been formed, the viscidity or ropiness of the latter impeding the crystalization of the remaining cane sugar. Vegetable juices containing fruit sugar possess a body capable of passing into a ferment, which at the proper temperature, induces the alcoholic fermentation in such juices.

Cane and fruit sugars are rarely employed by the vinegar maker. They generally command a higher price in their state of sugar, than when converted into acetic acid, and it is only in exceptional cases that they are thus employed. In the case of wine, cider, &c., used for the finer vinegars, and which ·arise from fruit sugar, it need scarcely be said, that these are made for beverages which command a higher price; the ultimate conversion into vinegar depending upon their souring, or upon other considerations, as of supply and demand, &c.

3. *Raisin Sugar.*—This sugar may be called, at least in our country, the source of all the vinegar made by the quick process, because all of our commercial alcohol arises from the fermentation of raisin sugar, formed in the brewing of grain of different kinds. This sugar bears two other names, *glucose* and *grape sugar*. Both of these designations are misnomers. Glucose means *sweet*, yet raisin sugar is inferior in sweetness to either cane or fruit sugar. It is estimated that one pound of cane sugar is equivalent in *sweetness* to from two to three pounds of raisin sugar. As to the other name, grape contains by far more

fruit than *raisin* sugar, although the former is transformed gradually into the latter when the grapes are dried and kept.

Raisin sugar appears to arise always from a change of fruit sugar, by the assumption of two atoms of water, ($C_{12}H_{12}O_{12} + 2HO = C_{12}H_{14}O_{14}$). The granular crusts which form in jars of preserved acid fruits, consist of raisin sugar. The cane sugar employed in making such preserves is converted into fruit sugar (?) by the acid of the fruit. This, by time and the absorption of water, changes to raisin sugar which crystalizes out.

Raisin sugar is more difficult of crystalization than cane sugar. It forms warty, cauliflower-like aggregations of grains. It is also less soluble, one pound requiring in the cold one and a half pounds of water for solution. On the other hand, it dissolves more freely in alcohol; one part of raisin sugar being dissolved by sixty parts of boiling absolute alcohol, and by from five to six parts of alcohol of 83 per cent.

The following is the action of heat upon this sugar:

At 140° Fah., it softens, and at the temperature of boiling water, it is completely liquid. At the latter temperature, it *loses two atoms of water* and becomes $C_{12}H_{12}O_{12}$, but this is not fruit sugar, for it acts differently upon polarized light. The solution of this changed sugar when evaporated, yields a pitchy mass when evaporated to dryness. Abandoned in contact with water for a time, this mass returns gradually to crystalizable raisin sugar.

At a higher temperature, raisin sugar like cane sugar, becomes *caramelized*. Raisin sugar combines less readily with bases than cane sugar, nor is it changed, like the latter, by the action of dilute acids.

Oil of vitriol dropped upon raisin sugar dissolves it without blackening, forming a chemical compound. Cane sugar under the same circumstances, yields a charred mass. Upon this difference is due the following characteristic test, which is said to indicate one millionth part of cane sugar in solution. Mix one part by weight of oil of vitriol with from five to seven of water. Add some of this acid to the suspected sugar solution, and expose to the temperature of boiling water. If cane sugar be present the liquid will turn dark.

A very striking characteristic of raisin sugar and one which enables us to distinguish it from cane sugar, and even to determine its quantity in solution, is its action upon salts of copper at the boiling temperature. Sulphate of copper, tartrate of potassa and caustic potassa are dissolved together in water, and the intensely blue liquid resulting is filtered. If weak solutions of sugar are to be tested, the solution should not be too blue; in such a case, dilute with water to the proper strength. If to a portion of this test, boiling, a solution containing raisin sugar be added, the test liquid will lose its blue color, and an orange red,

THE SUGARS.

granular solid, will separate from it. This precipitate, which is the suboxide of copper, results from the sugar depriving the copper test of a portion of its oxygen. Cane sugar does *not* and fruit sugar *does* produce the same result. If, however, a solution of cane sugar be boiled with a little oil of vitriol, it becomes converted into fruit sugar, which will give the above mentioned reaction with the copper test. A method of analysis for determing the actual amount of sugar in a solution, is founded upon this reaction with salts of copper, as follows. Take a measure capable of containing about two fluid ounces. As it is not necessary to know the exact capacity of this measure, it may be made from a glass stoppered bottle, by simply filing a small channel in the stopper, longitudinally. If this bottle be filled to the brim with any liquid and the stopper inserted, the excess of liquid will escape by the channel in the stopper, and thus the bottle always measures exactly the same quantity. Dilute a quantity of the copper test so that by repeated trial, it is found that ten grains of raisin sugar exactly decolorizes a *measure full* of it.

We are now prepared to determine in a few moments the quantity of raisin sugar in solution. Thus, boil one measure full of the copper test in a porcelain dish, and add the sugar solution gradually, from one of the graduated vessels, to be described on a future page; stop as soon as decoloration takes place, and read the number of parts of the saccharine liquid, that have been required to produce this effect. That number contains ten grains of raisin sugar. If we know the relation existing between the graduated measure and a gallon, we can, of course, by a simple rule of three calculation, ascertain how many grains of sugar are in a gallon of saccharine liquid. To apply this analysis to cane sugar, we must first boil the sugar solution with a little oil of vitriol, and having saturated the excess of acid by caustic potassa, proceed as in the former case.

Finally, a mixture of cane and raisin sugar may be thus analyzed. First, determine the decolorizing power of the mixture, which gives the amount of raisin sugar. Then, treat with acid and potash to transform the cane sugar, and determine the amount of decolorizing power which the saccharine solution has thus acquired. I have thus described an easy mode of analyzing the sugars, in case it should be required. Far simpler and more reliable, although more expensive at first on account of the apparatus, is the method of analysis by polarized light.

Preparation of Raisin Sugar, on a large scale.— The vinegar manufacturers generally employ alcohol already made. But in some instances they use potato starch or raisin sugar formed from the same, either by dextrine or by sulphuric acid. In all cases the sugar must be converted into alcohol by fermentation before it can be transformed into vinegar. This is effected in a small fermenting tun at the vinegar works.

In the following description, it will be seen how starch may be transformed upon a large scale into raisin sugar for the use of the vinegar maker by the action of oil of vitriol.

The operation is carried on in large tuns, which are two-thirds filled with the mixture, and in which the temperature may be raised to the boiling point by blowing in steam.

For every 100 pounds of starch from one to three pounds of oil of vitriol, and from 150 to 300 pounds (15 to 30 gallons) of water are taken. A portion of the water is used to dilute the acid, the rest to mingle with the starch. The diluted acid is introduced into the tun, and the steam turned on so as to bring it to the boiling temperature. The starch having been mingled with the water as stated, so that it may be poured in a thin stream, is now brought into the boiling acid in ten successive portions, and the temperature is maintained for 30 or 40 minutes, after the addition of the last portion. By this time the conversion into raisin sugar has been effected; but the fact is ascertained by suffering a few drops to cool upon a plate, and then adding a little solution of iodine, which strikes a blue color as long as any starch remain unchanged. In this operation a portion of the starch may be converted into gum, which is capable of being changed to sugar by a more prolonged boiling of the acid solution. This point is ascertained by filtering a small portion of the liquid, and adding then an equal bulk of strong alcohol. The presence of gum is indicated by a flocculent precipitate; gum being insoluble in alcohol.

The sugar being formed, the next step consists in removing the acid; which is effected by adding gradually, finely powdered limestone, or chalk rubbed up with water, to the liquid in the tun, until it ceases to redden blue litmus paper. The white sediment which thus forms is sulphate of lime or plaster of Paris. Every pound of oil of vitriol yields a pound and a half of this plaster; a small portion remains dissolved. After standing for twelve hours, the clear liquid is drawn off, and poured through bone black to decolorize it, after which it is boiled down to the strength proper for fermentation to alcohol.

If it be required to crystalize it, it must be boiled to a strength of 32° Beaumé, and set aside for eight days. One hundred pounds of *perfectly* dry starch may yield the same weight of dry raisin sugar. But since the apparently dry starch of commerce contains 18 per cent. of water, 100 pounds of such starch are capable of yielding not more than 82 pounds, or in practice, 80 pounds of raisin sugar. Hence it follows that we may obtain from 100 pounds of starch 800 pounds of syrup, of 10 per cent. saccharine strength, or proper for fermentation. The small amount of sulphate of lime which the syrup contains, does not injure it. The foregoing process may be improved by using quick lime for neutralizing the greater part of the acid. When limestone or chalk is employed, the escape of the carbonic acid is very troublesome, causing the liquid to boil over. Caution must be observed in the use of quick lime, as an excess injures the sugar. The liquid must become perfectly cold, and a half a pound of quick lime in the state of whitewash added very slowly for every pound of oil of vitriol present. A small portion of carbonate of lime is then sufficient to complete the neutralization.

The action of Sugar upon Polarized Light.—The effects of polarized light with the sugars are so characteristic of their different varieties, that a history of sugar would be imperfect without some allusion to such phenomena. I will therefore endeavor briefly and popularly to explain the phenomena.

To do this we must recall some of the properties and laws of light. Light always travels through a medium in right lines or rays. A ray of light reaching a body may be *either wholly or partially* reflected, refracted or absorbed.

If it leave the body at an equal angle, as a billiard ball rebounds from the cushion, it is "*reflected.*"

If it traverse a transparent body, more or less bent from its original course, (that is, making an angle with said course,) it is *refracted.*

If it strike a black body it is "*absorbed.*" If the body be colored, it is partially absorbed.

White light is, as is well known, composed of

the seven rainbow colors. Any one of these colors may be reflected refracted, or absorbed. The law of reflection is the same for every color, viz.: "the angle of reflection is equal to the angle of incidence."

Refraction, on the other hand, varies: 1st, with the transparent body; 2d, with the color of the ray; for first, white light is refracted to a different degree for different transparent bodies. The refraction is measured by a mathematical relation, which need not be given here. Secondly, the colored rays are refracted to a different degree by the same transparent body. It is this property of the colored rays which enables us to prove the compound nature of white light. Refraction has another peculiarity about it. If a ray of light strike a transparent body perpendicularly to a plane surface, it will be totally transmitted *without refraction*. If it strike at a certain angle peculiar to the body, it will be totally reflected *without* refraction. If it strike at an intermediate angle, part of the ray will be reflected, and the rest refracted, according to the nature of the transparent body. It concerns our purpose to observe that in the last case both the reflected portion and the refracted portion of the ray contain *polarized light*. Light may be polarized by reflection at certain angles from different substances, as water, black glass, &c.; by refraction, as by bundles of glass plates, carbonate of lime, and other crystals; and by transmission as through plates of the mineral tourmaline. By these means, a ray of white light, or a ray of any color, may be polarized. The following is the difference between common light and polarized light. If we permit a beam of any kind of light to enter a dark room through a circular aperture in one of the shutters, we will find that it may be reflected from, or transmitted by any crystalized or uncrystalized body in the very same manner, and with the same intensity, whether the surface of the body is held above or below the beam, on the right side or on the left, or on any other side of it, provided that it falls on the surface in the same manner.* In other words, a ray of common light has the same properties on all of its sides.

Now, if this same ray of light be permitted to pass through a crystal of Iceland spar, or to fall upon a plate of glass at the angle of 56°, the two rays into which it is divided will be *polarized*. They will have different properties upon different sides, as the magnet has different properties at its two poles. Take one of these rays of polarized light, and apply another plate of glass to it at the polarizing angle, (56°,) and rotate the glass, (keeping it always at the angle 56°,) and it will be found that on two sides, which are opposite, the ray will be reflected, while on two others, intermediate with the former, the ray will not be reflected at all.

We may also perceive this polarity by transmit-

* Sir. D. Brewster.

ting the ray of polarized light through another crystal of Iceland spar, or a plate of tourmaline, which we suffer to rotate on its axis. Beginning at the point where most light passes through the crystal, we will find on rotating that the light fades until it disappears at a quarter of a revolution, gradually reappearing to the original brightness at the half revolution, disappearing at three-quarters of the revolution, and appearing at the full revolution.

This is a very remarkable property of light, and one which has received an extended application in science and the arts. The body which polarizes the ray of light is called the "*polarizer,*" the body, by means of which we observe its polarity, is called the "*analyser.*" If we interpose between the polarizer and analyser, sections of doubly refracting crystals, there will be seen rings and curves of the most brilliant colors, caused by portions of the crystal of different molecular constitution refracting differently the white polarized ray, separating it into prismatic colors. In the same condition, granules of starch appear of brilliant white upon a dark back ground, each granule containing upon its surface a *black cross*.

Solutions of sugar have the property of rotating the plane of polarization to the right hand or to the left, a phenomenon called circular polarization.

The "plane of polarization" is that plane which contains the ray of light *before* and *after* its polarization. It is in this plane only that the polarized ray may be observed by the "analyser." As stated, two positions of the analyser, 180°, that is, half a circle apart, will permit the polarized ray to be seen by reflection or refraction, in which case the refracted or reflected ray is in the plane of polarization. In two other positions of the analyser, one quarter of a circle distant from the former two, there is total darkness. Polarize a ray of red light, and examine it with the analyser; in two positions of the analyser, it will be totally transmitted, and in two others totally shut off.

To illustrate circular polarization; using red light, place the polarizer and analyser for darkness. Then interpose a glass tube, of 8 inches in length, filled with a solution of sugar; the field of view is no longer dark, but red, and the analyser must be rotated a little before darkness is restored. The sugar has "*rotated*" the plane of polarization. With a solution of cane sugar, the analyser must be turned a little to the right before it darkens; with fruit sugar it must be turned to the left. The sugar has twisted the plane of polarization, and that is all; for if we rotate the analyser we shall find at 0°, perfect darkness; 90° perfect light; 180° perfect darkness; 270° perfect light; and intermediate shades of light between these points. If, in the above experiment with sugar we employ *white* light, the field of view will *never be dark*, but will change through the prismatic colors. The reason of this is that sugar rotates the plane of polarization to a different degree for every color of which white light is composed; consequently, in rotating the analyser, the maximum of light for one color coincides with more or less darkness for the other colors. In rotating the analyser to the right with a right polarizer, as cane sugar, the order of the colors is red, yellow, green, blue, violet, red. With a *left polarizer*, as fruit sugar, the analyser must be rotated to its *left*, to produce this order of colors.

In determining the degree to which the plane of polarization is rotated by different liquids, we must place them in tubes of the same length, since the thicker the stratum of liquid the greater is the rotation. Having placed the polarizer and analyser for darkness with white light, we insert the sugar tube and ascertain how many degrees the analyser must be turned to develop a peculiar violet shade, which is selected because the color changes rapidly to blue on the one hand, and red on the other, by rotating the analyser. At the same time we ascertain whether the substance is a right or a left handed polarizer. A method of analyzing sugar is founded upon these principles. By the law of circular polorization:

1st. For the same substance and for the same thickness of the stratum of liquid, the stronger the solution the greater is the rotation. This affords a means of analyzing sugars by comparing their solutions with a standard solution.

2d. In solutions of the same strength different bodies rotate the plane to a different degree. Thus dextrine rotates to the right to the greatest degree of all substances, whence its name "*right handed*." The rotation of cane and raisin sugar is also to the right, but to a less degree than that of dextrine.

The plane of polarization is rotated to

The left by	The right by
Fruit sugar,	Dextrine,
Gum Arabic.	Raisin sugar,
	Cane sugar,
	Milk sugar.

When fruit sugar becomes raisin sugar by long standing, it becomes right-handed. Cane sugar treated by an acid becomes left-handed fruit sugar.

When raisin sugar loses two atoms of water by heat, it acquires the same formula as fruit sugar; but it is not such, for it continues to be *right*-handed with respect to polarized light.

This difference of behaviour of the different sugars with polarized light is due to a constitutional difference of molecular arrangement, and proves that they are essentially different bodies, though closely related.

The sugar of the flower is cane sugar, which becomes the fruit sugar of honey by passing into the body of the bee. Honey exposed to the air for some time experiences another change from fruit sugar to raisin sugar.

Cane sugar, . . . $C_{12}H_{11}O_{11}$
$+HO=$Fruit sugar, . . . $C_{12}H_{12}O_{12}$
$+2HO=$Raisin sugar, . . $C_{12}H_{14}O_{14}$

Unfortunately the present state of chemistry does not enable us to go backward from raisin sugar to cane sugar. A fortune awaits the discoverer of a cheap process for effecting this reaction, cane sugar being much more valuable than either raisin or fruit sugar.

CHAPTER III.

ALCOHOL.

Since in the manufacture of vinegar, some employ wholly, or in part, a fermented liquid, which they prepare themselves, we may profitably in the present chapter, study the principles as well as the methods upon which the manufacture of alcohol depends.

A solution of pure sugar remains unchanged at all temperatures. If, however, we add a proper "ferment" at the temperature of 70° Fah., decomposition will set in and the molecule of sugar will be broken up into carbonic acid and alcohol. Ferments are certain bodies containing nitrogen, and undergoing decomposition. For example; when albumen, as in the white of egg, fibrine, as in the fibre of muscle, caseine, as in cheese, the gluten of seeds and vegetables, or other nitrogenized bodies of similar nature are exposed with water to the air, they do not delay to decompose. If, in this state, they be added to a solution of sugar at the summer temperature, the alcoholic fermentation takes place. The ferment called yeast is composed chiefly of vegetable egg-shaped cells, and if observed with the microscope in fermentable solutions, its growth by budding may be perceived. Whether this vegetation is a result of the fermentation, or whether the latter is a consequence of the vegetation, is yet a disputed point. Fresh yeast has the following composition before and after fermentation:

	PER CENTAGE.	
	BEFORE.	AFTER.
Carbon,	47.0	47.6
Hydrogen,	6.6	7.2
Nitrogen,	10.0	5.0
Oxygen, (about,)	35.0	

In brewing, yeast increases one-fourth of its original weight, by the aforesaid growth. This increase arises from a transformation of the glutenous or nitrogenized matter present in the fermenting liquids. When pure sugar solutions are fermented, no increase of yeast takes place. If the sugar be in excess, the yeast remains in an altered condition and inoperative to produce fermentation in another liquid.

The following is Liebig's theory of fermentation, which though not altogether satisfactory, is at least, the best we have. The ferment is undergoing a change, by reason of its decomposition excited by the oxygen of the air; consequently, *its atoms are in motion*, this motion is communicated to the sugar atoms with which it is in *contact*, so that they fall apart as carbonic acid and alcohol.

The following conditions are imperative for fermentation.

1st. The ferment is created by the exposure of certain nitrogenized bodies to the oxygen of the air. As soon as the ferment exists, fermentation takes place independently of the air. This fact has been proved by the following simple experiment. Fill with quicksilver a glass tube, so that no air bubbles remain attached to the side of the tube; close the end with the finger and insert it in a vessel filled with quicksilver. Ripe grapes may be so pressed under the mercury that their juice will rise into the tube, taking the place of a portion of the mercury. This juice not having come in contact with the air, will keep for an indefinite time. If now the smallest bubble of air be admitted to the juice, a ferment will be formed from the nitrogenized constituents of the juice, this will act upon the sugar and convert it into carbonic acid and alcohol. The gas may be seen collecting in the tube. The remaining liquid will be found to have lost its sweet taste and alcohol may be distilled from it. This is a beautiful and instructive experiment. The principle involved, induced the discovery by Appert of preserving fresh meats, fruits and vegetables by heat and hermetically sealing. Ripe fruits evidently keep as long as they do, because their juices are contained in separate cells, and the whole covered with a waxed skin, excluding perfectly the air. In Appert's process, which is now employed universally in the household, the vessel is filled with the fruit, vegetable or meat, and water at the boiling temperature. Having then been closed, it is kept for a short time at, or a little above, the temperature of boiling water. Any pre-existing ferment is thus destroyed. The small amount of oxygen present in the air in the interstices is gradually absorbed at the boiling temperature by the animal or vegetable contents of the can.

2d. Fermentation requires *contact* of the sugar solution with the ferment. Place some sugar dissolved in ten times its weight of water, in a wide mouthed bottle. Take a tube, open at both ends, and place it by means of a perforated cork (not air-tight), so that one end may dip in the sugar solution. Having cleansed this tube, tie filter paper tightly over its lower end, making a porous diaphragm. Place now some yeast in the tube, and insert the latter in the bottle. The yeast is thus separated from the sugar water by the porous paper, which will permit the passage of a liquid, but not that of a solid. Fermentation will *not take place in the bottle*, but what sugar water filters into the tube through the paper, will be fermented, proving conclusively that contact is necessary for fermentation.

3d. The following conditions arrest, modify or influence fermentation. The temperature is important. That most favorable to the alcoholic fermentation ranges between 68–77° Fah. At a low temperature, the fermentation is very slow. Bavarian beer is brewed between 32–46½° Fah.

In brewing malt liquors, a different kind of yeast is generated from the gluten of the malt at different temperatures. Thus, at the highest temperature the yeast floats and presents under the microscope the appearance already described. At low temperatures, the yeast, in the form of single egg shaped globules, sinks to the bottom of the fermenting tun. The limits for ordinary brewing are not lower than 51°, nor higher than 86°.

A boiling temperature at once arrests fermentation by destroying the ferment. The presence of too much sugar takes from the activity of fermentation. The most favorable strength is ten weights of water to one of sugar. Whatever destroys or removes the yeast arrests the fermentation. Thus filtering removes the yeast. The same is *killed* by certain essential oils, as that of mustard, sulphuric and sulphurous acid, the sulphites, &c. The following substances paralyze the ferment; much alcohol, common salt, cyanide of mercury, corrosive sublimate, pyroligneous acid, nitrate of silver, &c. Arsenious acid and tartar emetics, which are violent poisons to man, do not paralyze the fermenting action of yeast on sugar.

Such being the general principles of the alcoholic fermentation, it remains for us to inquire what *chemical* changes take place during the process. As regards the ferment, we are yet ignorant of the chemical changes to which it is subject; we only know that in a fermenting liquid, if albuminous matters are present, the ferment increases or grows.

The chemical transformation of the sugar is simple. One atom of sugar is split up into four atoms of carbonic acid, two atoms of alcohol, and in the case of raisin sugar, in addition, two atoms of water.

We have to consider the fermentation of three kinds of sugar, viz: that of, 1st. Cane sugar ($C_{12}H_{11}O_{11}$), which always before it ferments passes into 2d. Fruit sugar ($C_{12}H_{12}O_{12}$), and 3d. Raisin sugar ($C_{12}H_{14}O_{14}$), which loses two atoms of water with the greatest facility, passing into a body having the same composition as fruit sugar though not identical with it. As it differs not for the explanation of the alcoholic fermentation, whether the two atoms of water in raisin sugar are parted with *before* or *during* the process, let us assume that all sugar undergoing fermentation is composed of 12 atoms each, of carbon, hydrogen and oxygen. The change is then expressed thus,

Four atoms of carbonic acid,	C	—	O_2
	C	—	O_2
	C	—	O_2
	C	—	O_2
Two atoms of alcohol,	C_4	H_6	O_2
	C_4	H_6	O_2
Equal one atom of sugar,	C_{12}	H_{12}	O_{12}

ALCOHOL. 97

This result is also expressed by the following formula:

Sugar = Carbonic Acid + Alcohol
$C_{12}H_{12}O_{12}$ = $4CO_2$ + $2(C_4H_6O_2)$

We may readily learn from this formula, how much alcohol a given weight of sugar can yield. By substituting 6 for the carbon, multiplying it by the number attached, and proceeding in an analogous manner with hydrogen, whose equivalent is 1, and oxygen, whose equivalent is 8, we change the above expressions for atomic constitution into those of the combining weight of the several bodies. For example,

C_{12} = 6×12 = 72
H_{12} = 1×12 = 12
O_{12} = 8×12 = 96

Weight of 1 atom of sugar, = 180

C = 6
O_2 = 8×2 = 16

Weight of 1 atom of carbonic acid, 22
4

Weight of 4 atoms of carbonic acid, 88

C_4 = 6×4 = 24
H_6 = 1×6 = 6
O_2 = 2×8 = 16

Weight of 1 atom of alcohol, 46
2

Weight of 2 atoms of alcohol, 92

9

Hence 180 pounds of fruit sugar will yield 88 pounds of carbonic acid and 92 pounds of absolute alcohol. Consequently, by the rule of three, if 180 pounds of sugar give 92 pounds of alcohol : : 100 pounds of sugar : will yield $\frac{92 \times 100}{180} = 51 \cdot 12$ of alcohol, and we have the following per centage result:

> Alcohol, 51.12
> Carbonic acid, 48.88
> ———
> Sugar, 100.00

If we weigh the sugar as fruit sugar, we may say in round numbers, that sugar is capable of producing half its weight of alcohol.

If we weigh cane or raisin sugar, we must, for strict accuracy, modify this expression. Thus, cane sugar ($C_{12}H_{11}O_{11}$), contains 1 atom of water *less* than fruit sugar; hence, from the number 180 in the above expression, we must subtract 9, the weight of 1 atom of water ($HO = 1+8 = 9$), and the expression becomes, 100 pounds of cane sugar yields $\frac{92 \times 100}{171} = 53.22$ pounds of alcohol.

Raisin sugar, on the other hand, contains 2 atoms of water ($2 \times 9 = 18$), *more* than fruit sugar. Hence adding 18 to 180 we have, 100 pounds of raisin sugar yields $\frac{92 \times 100}{198} = 46.46$ pounds of absolute alcohol. This affords an illustration of the use and convenience of chemical formulæ.

Starch and gum have the same formula, $C_{12}H_{10}O_{10}$, and are converted into fruit sugar by the assumption of two atoms of water—162 pounds of gum will yield 180 pounds of the said sugar, which, in its turn, will produce 92 pounds of absolute alcohol. If 162 pounds of starch give 92 pounds of alcohol : : 100 pounds will yield 56·79 of alcohol. This yield of alcohol is, however, never practically obtained in brewing, because a portion of the starch is converted into a *permanent* gum incapable of the saccharine transformation. As the formation of permanent gum is influenced by the temperature during the mashing, (see Ure's Dic., Art. Beer,) it becomes important for the vinegar maker and distiller who brew with a view to alcohol and not to beer, to be well acquainted with the principles governing this formation.

Let us proceed to the practical application of the fermentative process. The simplest case is wine. If the juice of the grape or of any other saccharine fruit or vegetable be exposed to the air, certain nitrogenized compounds existing therein become ferments, which convert the sugar into alcohol and carbonic acid. Thus, grapes yield wine, apples cider, pears, perry; the sap of certain palms, toddy; of the sugar cane, guarapo, or sugar cane wine; of the American aloe, pulque, &c., &c. These wines, distilled, yield spirituous drinks containing a larger proportion of alcohol; thus, brandy from grape wine; kirsch-wasser from fermented cherry juice; rum from sugar cane, wine or fermented molasses;

aguardiente from pulque, &c. To these may be added arrack from koumiss, or fermented mare's milk, and whisky, from fermented grain.

The acidity or sweetness of the different wines, depends upon the relative proportion of the ferment to the sugar; upon the saccharine strength of the vegetable juice, and upon the method of the manufacture generally.

Johnston* gives the following tables of the relative sweetness and acidity of the respective well known wines:

TABLE OF SWEETNESS.

Claret, Burgundy, Rhine and Moselle,	Wines contain no sensible portion of sugar.	
Sherry,	from 4 to 20 grains sugar per ounce	
Madeira,	" 6 to 20	" "
Champagne,	" 6 to 28	" "
Port,	" 16 to 34	" "
Malmsey,	" 56 to 66	" "
Tokay,	74	" "
Samos,	88	" "
Paxarette,	94	" "

SCALE OF ACIDITY.

Sherry the least acid.
Port next "
Champagne " "
Claret " "
Madeira " "
Burgundy " "
Rhine Wine " "
Moselle the most acid.

* Chemistry of Common Life.

ALCOHOL. 101

The acidity of wines is due to Tataric acid in combination as bitartrate of potassa from the grape juice, and partially to acetic acid, arising from a transformation of a portion of the alcohol.

If the grapes from which the wine is made were unripe, citric acid is present in the beverage.

The amount of alcohol varies not only with the different kinds, but in different varieties of the same wine. This alcoholic strength depends upon the kind of grape used, upon soil, climate, season, culture, and method of manufacture, and upon the mode of storing the wine.

Sparkling or effervescent wines are a variety bottled before the fermentation is completed; by which a portion of the carbonic acid of fermentation is imprisoned in the wine and kept in solution by its own pressure, and escapeing when the vessel is uncorked.

Brande gives the following table, in which the per centage (by measure) of absolute alcohol is stated for several well known wines:

TABLE.

100 *measures of the wine, at 60° Fah., contain the following measures of absolute alcohol.*

Port wine,	19·82
"	23·92
Madeira,	17·91
"	22·61
Sherry,	17·00
"	18·37
Claret, (Bordeaux,)	11·95
" "	15·11

9*

VINEGAR MANUFACTURE.

Lisbon,	17·45
Malaga,	15.98
Malmsey,	15.91
Marsala,	14·31
"	15·98
Champagne, (rose,)	10·46
" (white,)	11·84
Burgundy,	13·34
"	11·06
Hermitage, (white,)	16·14
" (red,)	11·40
Hock,	13·31
"	8·00
Vin de Grave,	11·84
Frontignac,	11·84
Cape Madeira,	16·77
Muscat,	17·00
Constantia,	18·29
Tokay,	9·15
Lachrymæ Christi,	18·24
Currant Wine,	19·03
Gooseberry Wine,	10·96
Elder Wine, ⎫ Cider, ⎬ Perry, ⎭	9·14
Brown Stout,	6·30
Ale,	8·00
Porter,	3·89
Rum,	49·71
Hollands,	47·77
Whisky, (Scotch,)	50·20
" (Irish,)	49·91

The foregoing tables, used in connection with others to be given in this work, are useful in calculating the acid strength of vinegars capable of being manufactured from the respective wines, such acid strength depending of course upon the per centage of alcohol in the wine.

Besides the ingredients already enumerated, wine contains coloring, organic and mineral matter, the latter derived from the soil. These of course are found in the vinegar manufactured from such wines.

Distilled wines (spirits) do not contain the mineral salts and extractive matter. The slight color which some of them possess is due probably to the action of heat upon the extractive matter of the wine during the process of distillation giving rise to volatile coloring matter in small quantity.

A very important class of ingredients, not only in wines but spirits, comprehends certain volatile aromatic liquids, existing in very minute proportions. To these are due the aroma and bouquet of wines, giving them their different and characteristic flavors, and influencing so strongly their relative values. These substances are formed by the fermentative act from substances existing in the juices from which the wines are made. By the acetic transformation they undergo changes together with the alcohol, communicating certain flavors to the resulting vinegars. By reason of their presence, the vinegar from wine stands pre-eminent above all others. A greater chemical knowledge with respect to them is increasing daily, and hopes are entertained thereby of greatly improving the quick vinegar manufacture which employs pure spirits.

The second class of fermented liquids embraces all cases in which sugar has been employed in a pure state, and in which a *ferment* has to be *added* to the saccharine solution.

Any sugar mentioned in Chapter II., may be dissolved in ten times its weight of water, the solution brought to a temperature 68°–77° Fah., and good brewers' yeast added. For every 100 pounds of sugar, 1½ pounds of yeast (estimated in the dry state) will be required. Fermentation takes place rapidly, especially in the case of fruit and raisin sugars. Cane sugar requires a longer time, and passes before fermentation into fruit or an analogous sugar, as has been shown by experiments with polarized light.

The curious fact has been proved that this change is due to a vegetable acid or acids present in the yeast, and as this acid is of vegetable nature, a longer time is required to transform cane to fruit sugar, than in our experiments with mineral acids. By Regnault's authority, it is stated, that yeast freed from its acid by washing, *will not ferment* solutions of cane sugar until by exposure to the air a fresh quantity of acid is generated in the yeast by its decomposition.

The strength of a pure solution of sugar may be ascertained by simple inspection with the saccharometer; the advance of fermentation may be watched by noting the diminution of the specific gravity of the liquid; and the amount of alcohol present can be determined by the alcoholometer.

These tests and the necessary instruments will be explained upon a future page.

The third class of fermented liquids embraces all cases of grain fermentation, and that of analogous substances, in which starch is converted into raisin sugar by the action of diastase formed during the operation, while sugar is converted into alcohol by a ferment also generated during the operation. The process is called brewing. The brewer and the distiller employ processes *generally* the same, the different nature of the product required by each, involving a slight difference of treatment of the materials employed.

The brewer strives after the finest flavored beer or ale, and one which may be readily preserved; while the distiller regards less the taste of the fermented liquid than the amount of alcohol which he is able to form at the same cost, and he is prepared to at once distill off this alcohol.

The fermentation of bread involves some of the principles of this class of fermented substances. To flour, kneeded with water, is added yeast, or leaven, which is fermented dough, and the resulting mass is exposed to a warm temperature. By action of the ferment, the starch of the flour is transformed into sugar and gum, from portions of which alcohol and carbonic acid are generated by fermentation; while from the gluten of the flour, by the generation of more ferment, the process is accelerated. As dough is of a tenacious character, (owing to the gluten,) the carbonic acid is impeded in its escape, and puffs up the dough, forming a porous mass. Baking arrests the fermentation, renders the mass more porous by the expansive action of heat upon the carbonic acid and water, drives off carbonic acid and alcohol, and renders the mass more solid by the evaporation of a portion of the water, and by the coagulation of the albuminous matter by heat. In some of the large bakeries of Europe, the alcohol has been collected by appropriate devises, but the process has not been sufficiently remunerative to warrant its continuance.

THE ART OF BREWING.

Let us consider now in greater detail the art of brewing; a knowledge of which is of great importance to the vinegar manufacturer; for in many instances it is very remunerative to employ in this manufacture the starch of potatoes, or of some kind of grain which must be fermented before it is available. Besides, many have an erroneous impression as to the part which sugar plays in the vinegar process, which, if corrected, will enable the manufacture of a superior article at an inferior cost.

The art of brewing falls naturally into four stages. 1st, malting; 2d, mashing and preparing the wort; 3d, fermenting the same; and 4th, ripening and preserving the fermented liquid. The distiller, who brews grain for the alcohol, which he at once distills off, proceeds differently from the beer and ale brewer, in 2d, 3d, and 4th.

1st. In both processes the malting is similar, the object being the artificial germination of the grain. Here diastase is formed, which acts upon the starch, converting a portion of it into sugar and gum. When the process is sufficiently advanced, it is arrested by drying the grain, which is then "malt."

2d. Mashing is the preparation of a solution in hot water of the malt, and the further action of heat upon this solution, to which farinaceous substances have been added, the resulting liquid being called the "*wort.*" During this process, the sugar and gum are dissolved from the malt, and its diastase completes its action to convert the remaining starch into sugar, and to effect the same transformation upon the starch of the added farinaceous matters. This process is also arrested by elevating the temperature.

3d. In fermentation the sugar is converted into alcohol by the action of an added ferment, yeast. In beer brewing, a portion of the sugar remains after fermentation.

4th. The preservation of the beer and its pleasant taste are secured by hops, added during the preparation of the wort, by a secondary fermentation, which places carbonic acid in the beverage, and by the general manner, according to which the whole process has been carried on.

The distiller aims to prepare a solution of alcohol of definite strength at the least cost of time and fuel, and with the greatest freedom from fusil oil, which is a liquid of disagreeable taste, existing more or less in all fermented liquids, and supposed to be derived from the husks or skins of the grain, fruit, or vegetable. It is not looked upon with disfavor by the vinegar maker, as it changes by his process into aromatic substances, which improve the flavor of the vineger.

The distiller endeavors to manage the preparation of his worts so that it contain the greatest possible amount of sugar and the least unconvertible gum. In fermenting the wort he aims to leave *no sugar* unconverted into alcohol. In other words, he manages to convert as much starch as possible into sugar, and then alcohol; and does not care for the keeping properties of the resulting liquid, as he at once distills off the alcohol. Both beer brewers and distillers avoid all tendency to *acetification* in their process. This the vinegar brewer encourages, proceeding in other respects like the distiller.

Having thus taken a bird's eye view of brewing, let us consider it in greater detail, and especially with an eye to the vinegar manufacture.

I. MALTING.—The operation of malting is a beautiful illustration of the power which man possesses over the functions of vegetable life, making them subservient to his own life, well-being, and happiness. Starch is to be changed into alcohol; it must first pass into sugar, and *diastase*, a substance formed in the first development of the plant from the seed, must be had as a simple means of effecting the required transformation. The maltser places the seed in the condition of its natural development; in other words, to borrow a figure from the poultry yard, he *hatches* it. He stimulates its vital energy, until sufficient diastase is formed to effect his purpose; then puts it to a violent death to prevent a waste of starch. If the outer husk of a grain of barley or of other similar seed be removed, we shall find an envelope of hard cells (gluten) in close contact with each other. Enclosed in this shell is the starch, and at one end of the seed we shall find the germ of the future plant, destined to feed upon the starch and gluten until its rootlets penetrate the earth downward, and a stem and leaves elevate themselves into the air, to draw therefrom its gaseous food, carbonic acid. The development of seed to plant takes place *naturally*, when the spring rains have saturated it with moisture, and when the warmth of the approaching summer is beginning to be felt. When the conditions of moisture and warmth are absent, the seed refuses to germinate, its power in this respect remaining latent for an indefinite period. We have witness of this in the grains of wheat found in the sarcophagus of an Egyptian mummy, which, when planted, germinated, giving plant and seeds, from which arose a variety new to modern times, and which bears the name of mummy wheat.

Malting may be performed upon any grain; but barley is peculiarly suitable for the purpose from the quantity of diastase it gives rise to. The operation consists of three parts.

1st, steeping; 2d, couching; and 3d, drying.

1. *Steeping* is effected with water in stone or wooden cisterns, furnished with a large faucet and perforated plate to prevent the egress of the grain when the steeping water is drawn off. Enough water having been introduced into the cistern to cover the grain to a depth of six inches, (to allow for swelling,) the barley is added and stirred about with rakes. The imperfect grains which float are removed. If the barley be dirty, or should acetification begin, the water is renewed. During the steep, carbonic acid is formed from the grain, and held in solution together with extractive matter from the husks, which communicates to the steep water a yellowish color. The grain takes up one half its weight of water, and increases one-fifth in size, is lighter in weight, (if dried,) and paler in color by the loss of extractive matter. The maltster judges that the steeping is complete, when the grain may be readily pierced with a needle, and when a grain, upon being strongly pressed between thumb and finger, sheds its starch. If it remain in the husk, it has not been sufficiently steeped; if it exude a milky juice, it has been spoiled for germination by too long steeping. The time required for steeping depends upon the kind of grain, its age, and

the temperature of the water. Old grain requires longer steeping than new. For dry sound grain from 36 to 48 hours are required in summer, and in cold seasons from three to five days. At the completion of the steep the water is drawn off, fresh water added to wash the grain, and the latter suffered to drain off for several hours through the open faucet. The barley is now ready for the operation of

2. *Couching.*—At this stage, the germination of the grain is effected. A rather low temperature (not to exceed 62° in summer,) darkness and air are required to effect the object of couching.

The room is by preference a cellar, with a dry floor of stone, cement, or brick, having every crack carefully cemented. The windows are furnished with slats to admit air and exclude light. The barley is formed into a square heap, called the couch, of from five to sixteen inches in height, being somewhat higher at the edges, where the evaporation is greater than in the middle of the heap. The management of the couch requires and exhibits all the skill of the maltster, whose object is to cause every grain to germinate in like degree, and to arrest the germination at the proper point. He must keep the grain from drying, and prevent the temperature from becoming higher in one portion of the heap than in another, conducting the operation slowly, so that the rootlets are developed rather than the germ of the future plant. These purposes are effected by repeated movement of the couch, shovelling it into new heaps so that the intermediate grains become top ones; by altering the height of the couch to regulate the evaporation, and by managing the light to effect the manner of vegetation. In handling the couch wooden shovels are employed, care being taken not to break the grain, as such injured seeds could not germinate, but would decompose, giving a bad malt.

A badly managed couch would yield seeds fully sprouted; some undersprouted, which would not contain the desired diastase, some oversprouted, which would involve a loss of starch, and thereby alcohol, as the rootlets and acrospire* absorb starch, and are not converted into alcohol by the brewing process.

Germination is an oxidation process, a slow combustion. The seed absorbs oxygen from the air, and returns carbonic acid, giving rise to an elevation of temperature which is controlled by the maltster by altering the height of his couch.

At first no dampness is imparted to the hand when thrust into the heap. After a while a fruity smell is perceived, the temperature of the inside of the couch rises 10° above that of the atmosphere, and the hand thrust into the grain is bedewed with moisture. This stage of the operation is called sweating, and it is now that the germination begins. The fibrils of the roots be-

* The germ or portion of the sprouted seed which becomes stalk and leaves is called in brewing the "*acrospire.*"

gin to appear at one end of the seed, and shortly after an elevation is perceived at the same end. This is the acrospire, which in barley proceeds *under the husk* to the other end of the seed, where, if the process were not arrested, it would break forth to become the "*plumula,*" or stalk and leaves.

After the sweating process has continued for a short time, the couch is turned to bring about the same process for the external grains, and when the germination is fully started, the couch is lowered in height at every shovelling, until from 16 inches it becomes 3 or 4. Two turnings a day are customary.

By this means the too speedy germination of the seed is arrested. It must be remembered that the rootlets and acrospire, and especially the latter, destroy starch. If the temperature be kept down, and the light shut out, the acrospire does not develop itself well, and at the close of the process has not advanced much beyond half the length of the seed, while the rootlets have pushed to $1\frac{1}{2}$ times that length, and are curved, so that the seeds hook together. Malting of course involves a loss of some starch, without which the diastase could not be formed; scientific malting seeks to make this loss as small as possible.

At the close of couching, the barley, now well germinated, is dried quickly to arrest further development. This effect is produced either in the air or by malt kilns.

3d. *Drying.*—Kiln drying is effected at temperatures differing according to the color required for the malt, which is thus adapted to different varieties of beer. It must not be forgotten that an elevated temperature converts starch into permanent gum which does not ferment into alcohol, and hence for the vinegar maker and for the distiller, the lower the temperature of the kiln, the better is the malt. One hundred degrees, Fah., is the limit for their purpose.

By drying, the rootlets and acrospire become brittle, fall off and are sifted away from the "*malt.*"

One hundred pounds of barley, judiciously malted, weighs after drying and sifting, eighty pounds. The loss is as follows,

Malt,	80.0
Water in the barley,	12.0
Extractive matter removed by steeping,	1.5
Dissipated in the kiln,.	3.0
Loss by falling of the fibrils,	3.0
Waste,	0.5
Weight of original barley,	100.00*

Good malt exceeds the *bulk* of the original barley by from 8 to 9 per cent.

Ure gives the following characteristics of good malt. "The grain is round and full, breaks freely between the teeth, has a sweetish taste, an agreeable smell, and is full of soft flour from end to end. It affords no unpleasant flavor on being

---
* Ure.

chewed, and is not hard, so that when drawn across the fibres of an oaken table it leaves a white streak like chalk. It swims upon water, whereas unmalted barley sinks."

II. PREPARATION OF WORTS, OR MASHING.— The operation of mashing has for its object the transformation of starch to sugar by the aid of the diastase of malt. The result is a saccharine solution suitable for fermentation, which is called the "wort" or "worts." In malt itself a portion of starch has already submitted to a saccharine change by the diastase, and another portion has been changed to gum by the drying process, especially if the temperature of drying has been too much elevated.

In the preparation of fine ales, good barley malt is alone submitted to the mashing process; but the distiller and vinegar maker always add a certain quantity of unmalted grain, because the diastase in the malt is capable of transforming a much larger quantity of starch than exists in the malt. One part of pure diastase is capable of saccharifying 2,000 parts of pure dry starch. The relative proportion of diastase and starch existing in malt, depends upon the grain from which the malt is made, as well as upon the skill of the maltser. From ten to twenty-five pounds of finely ground malt will, with four hundred pounds of water at a temperature between 140–167°, convert into sugar one hundred pounds of pure starch. Any of the cereal grains, or the starch from potatoes, may be mixed with malt for mashing. A mixture of several sorts of grain is considered advisable, wheat with barley and oats; barley with rye and wheat, &c. One object of the mixture is to obtain a more porous mass for mashing by reason of the husks of the grain.

The relative proportion of raw grain and malt employed vary with the different manufacturers. Otto advises the vinegar maker to use equal parts of raw grain and malt. The grain and malt must both be ground, but not too finely, for not only would a pasty, impenetrable mass result, but it would be difficult to obtain a clear wort.

This comminution is attained by crushing with rollers, (which is preferable,) or by grinding between mill stones, or under edge stones, so that the hull of the grain is torn asunder and the pulverized starch set free. It is necessary to perform this operation a short time before preparing the mash, as the pulverized malt and grain spoil, especially in damp localities.

The mashing tun is a large wooden vessel, having a perforated false bottom a few inches above the real bottom. These perforations are so small that the crushed malt cannot pass, and are conical with the greater diameter downward to prevent choking. The faucet is, of course, below the false bottom. The amount of warm or cold water necessary to form a thin paste, having been introduced into the mashing tun, the mixture of malt and raw grain is added, and the mass thoroughly stirred by blades, revolving by hand power or by machinery, so as to obtain a uniform mixture, free from lumps. The proper amount of water, which has been heated in a large copper vessel, is now added, mixed with the mass, and the same suffered to stand with the tun covered from an hour to an hour and a half. An addition of a quart of skimmed milk to every one hundred pounds of malt and grain, is considered by Balling of advantage. At the expiration of the allotted time, the saccharification is complete. We can assume that practically in mashing, every pound of starch gives rise to a pound of mixed sugar and gum, and the sugar exceeds in proportion the gum the nearer the heat has been kept to the minimum temperature, 140° Fah. As the gum does not yield alcohol in this process, it becomes important to regulate the temperature in mashing. The heat should rise gradually and not exceed 151° Fah. This end is attained by adding the hot water gradually to the mixture, suffering it to flow upward through the false bottom, the mass being continually stirred. When the first wort is ready, the faucet should be opened and the wort returned to the tun until it flows clear. It is then drawn off as closely as possible. More water is added, stirred into the mass, and withdrawn in like manner, to obtain the saccharine solution with which the grain is saturated. A third quantity of water is *sprinkled* upon the mass in the tun and in passing through it displaces whatever valuable wort might otherwise remain. Further exhaustion of the mass is useless, for not only would a wort result too weak for profitable fermentation, but the value of the grain residue for feeding stock would be impaired. The worts when sufficiently cool, may be fermented; but a very important point with respect to their strength remains to be considered. This subject is called,

The concentration of the wort.—The alcoholic strength of the fermented liquid depends upon this concentration, which is governed by the quantity of water added during the mashing, less what evaporates during the subsequent boiling and cooling of the worts. How much water, therefore, must be added during the mashing process, is a question to be decided by the required strength of the fermented liquid.

With a perfect fermentation one pound of alcohol arises from two pounds of sugar. Every pound of alcohol yields a little less than $1\frac{1}{3}$ pounds of radical vinegar, that is, hydrated acetic acid, or the strongest possible vinegar. If then we seek to obtain a vinegar of given strength, we must bear in mind these data in order to obtain the proper concentration for the wort. Thus, a 12 per cent. solution of sugar will give a 6 per cent. (by weight) alcohol, which will yield a vinegar containing a little less than 8 per cent. (by weight) of hydrated acetic acid.

The residue obtained by the evaporation of

wort is called "malt extract." It contains sugar and gum together with a little coloring matter.

The following are Balling's results from the analysis of malt extract obtained from the different grains.

	Average percentage of malt extract.
From wheat or maize,	70
" barley,	60
" barley malt, (unkilned,)	57
" equal parts wheat and barley,	65
" equal parts raw and malted barley,	58

These numbers represent also the percentage of starch in the respective grains employed, since every pound of starch yields a pound of malt extract. Of course in the practice of brewing, a smaller proportion than the above of malt extract is obtained, since a portion remains in the grain residue, and in the last weak wort, with which it is saturated. For example, if the malt mixture contain 60 per cent. of starch, 100 pounds would be capable of giving 600 pounds, that is, 72 gallons of wort of 10 per cent. saccharine strength; but for the reasons assigned we are able to obtain in practice only 65 gallons of wort of 10 per cent. malt extract strength. To prepare a wort of 10 per cent. with the malt mixture generally employed, we must take for every 100 pounds, 84 gallons of water, of which we may use 24 gallons of temperature, 143°, to mingle with the grain, and 24 gallons at the boiling temperature, to effect the conversion of starch to sugar. This will leave 36 gallons to employ in two portions for washing out the wort from the residue of grain. The first wort flows off with a concentration of from 12 to 13 per cent.*

We are not, however, left to mere routine in this matter, since the saccharometer enables us always to test the strength of worts. The greater the percentage of malt extract in the worts the greater is its specific gravity, which the saccharometer exhibits by a simple inspection. This instrument is of the same principle as the alcoholometer, and the family name of all such is the hydrometer. For a description of them see a subsequent page.

In practice the worts are cooled to the temperature for which the saccharometer is graduated, and which is generally marked upon the instrument. The worts are placed in a tall glass cylinder, the saccharometer suffered to float therein, and at the water level upon the stem of the instrument, the strength of the worts may be read.

The strength of worts may also be inferred from its specific gravity, to be determined as will be described. If the specific gravity of the wort is taken at 63° Fah., its percentage strength of malt extract may be obtained by consulting the following table:

* Otto.

THE ART OF BREWING. 121

TABLE.

The specific gravity of saccharometer degrees at temperature 63° Fah.

Sugar in 100 parts of the solution.	Specific gravity of solution.	Sugar in 100 parts of the solution.	Specific gravity of the solution.
0	1·0000	39	1·1743
1	1·0040	40	1·1794
2	1·0080	41	1·1846
3	1·0120	42	1·1898
4	1·0160	43	1·1951
5	1·0200	44	1·2004
6	1·0240	45	1·2057
7	1·0281	46	1·2111
8	1·0322	47	1·2165
9	1·0363	48	1·2219
10	1·0404	49	1·2274
11	1·0446	50	1·2329
12	1·0488	51	1·2385
13	1·0530	52	1·2441
14	1·0572	53	1·2497
15	1·0614	54	1·2553
16	1·0657	55	1·2610
17	1·0700	56	1·2667
18	1·0744	57	1·2725
19	1·0788	58	1·2783
20	1·0832	59	1·2841
21	1·0877	60	1·2900
22	1·0922	61	1·2959
23	1·0967	62	1·3019
24	1·1013	63	1·3079
25	1·1059	64	1·3139
26	1·1106	65	1·3190
27	1·1153	66	1·3260
28	1·1200	67	1·3321
29	1·1247	68	1·3383
30	1·1295	69	1·3445
31	1·1343	70	1·3507
32	1·1391	71	1·3570
33	1·1440	72	1·3633
34	1·1490	73	1·3696
35	1·1540	74	1·3760
36	1·1590	75	1·3824
37	1·1641	75·35	1·3847
38	1·1692		

Let us return now to the final treatment of the wort, which we left separated from the exhausted malt residue. Beside sugar, gum, and coloring matter, it contains in solution several nitrogenized substances, as vegetable albumen and changed and unaltered diastase. It may either be at once cooled down to the temperature of fermentation, or boiled for a short time to separate by coagulation, the above mentioned nitrogenized matters. It is true that the presence of these renders the subsequent fermentation easier, and if the manufacture of yeast for sale is connected with the brewing, increases the quantity of this product. At the same time they render the vinegar made from such worts more liable to spoil and acquire a bad smell if the subsequent operations of fermentation and acetification are not carried on very slowly.

The worts are boiled in a large copper kettle, in which the whole operation of mashing is performed in some factories, having care to use stirring machinery to prevent the mash scorching upon the bottom of the copper. A portion of the albumen coagulates in clots, which are skimmed from the boiling liquid and placed in a small vessel to drain off the adherent wort. The rest of the coagulum is separated after the boiling by passing the wort through a sieve, or a strainer made from a basket lined with plenty of straw.

The clear wort is then drawn off into coolers, which are large shallow vessels, exposed to air currents at a depth of two or three inches. In warm weather, a coil of tube in which flows a current of cold water, aids in this refrigeration. It is important to cool the worts as rapidly as possible, and not to ferment them at a high temperature. If the fermentation commences in the coolers at a high temperature, the alcohol is partially, as it forms, converted into acetic acid. The fermentation should confine itself to the sugar and take place slowly to give the yeast time to separate. Acetic acid will hinder the perfect separation of these nitrogenized products, which will be found in the vinegar made from the fermented liquor, and as said before, render it more liable to spoil by keeping.

III. FERMENTATION.—The worts cooled to the proper temperature are ready for fermentation. This is affected in a vessel called the "*gyle tun,*" which must be large enough to allow for the rising of the yeast. When the temperature of the room containing the gyle tun is 60° Fah., the worts should be cooled to about 65°. For every 100 gallons of worts, from $\frac{1}{2}$ to $\frac{3}{4}$ of a gallon of good fluid yeast is added and well stirred in. Some prefer not to add the whole of the yeast at once; but to mix it with several gallons of warm wort, in which a very rapid fermentation is excited, and to add this mixture to the worts at once or at intervals. Balling advises the addition of a little malt meal or cold extract of malt to the fermenting wort, supposing that its diastase converts to sugar a portion of the gum in the wort. The length of time required for the fermentation depends upon the general manner of carrying on the process in which the proper regulation of the temperature plays a prominent part. After a few hours, a wreath of foam occasioned by bubbles of evolved carbonic acid, makes its appearance around the edge of the tun. As it widens, the temperature increases, and it at last overspreads the surface of the liquid and thickening, forms a high mass of foam which effectually excludes the air. The greater part of the yeast floats in this foam, although a portion falls to the bottom of the tun. The elevation of temperature and evolution of carbonic acid gas keep up a continual motion of the particles of the worts, which gradually exchange their sweet taste for an alcoholic flavor. The density diminishes from two causes; 1st, they lose sugar, the solution of which is heavier than water; and 2d, they gain alcohol, a fluid lighter than water. The intestine commotion at length begins to subside, the cover of yeast and froth becomes harder, brown in color, and loosens itself from the sides of the tun, and the temperature of the liquid falls to that of the surrounding air. These are all signs of the completion of the first stage of fermentation; but the brewer watches especially the lowering of specific gravity, or in his language the attenuation of the worts. I shall have occasion to explain more fully the subject of attenuation at the close of the present chapter, when discussing the alcoholic strength of the malt wine.

Upon the completion of the gyle tun fermentation, the top yeast is removed, the bottom yeast well stirred up, and the malt wine placed in large casks with open bungs, to purge itself. A slow after-fermentation sets in, and as the casks are kept constantly full of liquid, the remaining yeast escapes from the bung-hole. As soon as no more issues, the bung-holes are cleansed, and the bungs driven. A quiet fermentation establishes itself, which perfects the malt wine, and leaves it ready for the vinegar maker. I have said nothing about hops, which are not used by the vinegar manufacturer.

This operation of brewing for vinegar may be carried on with a much smaller capital than for beer brewing or distilling, especially where malt may be purchased.

The brewing process thus described may serve as a pattern for the conversion of any starchy or saccharine substance to alcohol. If we employ potatoe or other starch, we must first change it to sugar by malt by the operation of mashing, or by the action of sulphuric acid, as described in a former chapter. For raisin sugar, honey, &c., it is only requisite to make, by the aid of the saccharometer, a solution of definite strength, remembering that half the saccharometer degrees will give the percentage by weight of alcohol in the fermented liquid. Thus a solution indicating

12 per cent. by the saccharometer, will yield an alcohol of 6 per cent. by weight absolute alcohol, since every pound of sugar is capable of affording about half a pound of absolute alcohol. Having a saccharine solution of known strength, it remains to ferment it by the addition of yeast at the proper temperature as described.

PROPERTIES AND TESTS OF ALCOHOLIC SOLUTION.

The vinegar maker employs two kinds of alcoholic liquids; 1st, pure alcohol and water, or "spirits;" 2d, wines, cider, perry, malt wines, &c., which consist of alcohol water, gum, aromatic ethers, and a small proportion of mineral salts, derived from the earth by the flow of the sap. It is very important to learn the true alcoholic strength of any fermented liquor or spirits employed by the vinegar maker, and to be able to obtain this information for oneself, not taking it at second hand from interested parties. To give the different methods by which this end may be readily accomplished will be the object of the remainder of the present chapter.

When we say that a certain liquid contains so much per cent. of alcohol, we always understand "*absolute*" alcohol, that is, alcohol in its highest possible state of concentration. In speaking of the percentage of alcohol, we always mean, unless it is stated otherwise, percentage by volume; for example, 90 per cent. alcohol is that of which 100 gallons contain 90 gallons of absolute alcohol. If the percentage is expressed by weight, then in

100 *pounds* of, say 60 per cent. alcohol, there would be 60 pounds of absolute alcohol. I shall give tables, from which weight per cents of alcohol may be converted into volume per cents, and vice-versa.

Absolute alcohol contains 4 atoms of carbon, 6 of hydrogen, and 2 of oxygen. The specific gravity at 60° Fah. is 0·793. When the barometer stands at 30 inches it boils at 173° Fah. From this we may infer that the weaker a mixture of alcohol and water is, the nearer will be the approach of its specific gravity to 1, and of its boiling point to 212° Fah. The distiller obtains strong spirits by successive distillations. During the first distillation, the first products are the strongest in alcohol; and the boiling point of the liquid in the still rises gradually. By paying attention to this fact, the distiller is able by the first distillation to obtain several liquids of different but well known alcoholic strength. These are called technically "whiskys" or "high wines." The name whisky in this country is generally given to products arising from the fermentation of a mixture of grain in which the rye predominates; while in high wines, the maize exceeds the rye in the mashing process.

Towards the close of the distillation the *fusel oil* comes over in greater abundance than at the commencement. This unwelcome liquid is found in larger quantity in spirits distilled from potato or maize than from rye, barley, &c. It is not objectionable to the vinegar maker, because it is transformed in his process into aromatic ethers, which impart a fine flavor to his product. It is removed to a great extent from spirits by filtration through thick layers of charcoal, a process called technically, but improperly, "*rectifying.*"

Rectifying properly speaking consists in submitting the first portions of the first distillation to another distillation, in which we obtain first products of higher alcoholic per centage.

No number of rectifications will remove all of the water from spirits. The strongest rectified spirits of wine contain from 10 to 14 per cent of water, which is in the proportion of one atom of water to one atom of absolute alcohol. The last portion of water may, however, be removed by chloride of calcium, or by quick lime, substances having a great affinity for water and not injuring alcohol. The following is the process: Quicklime is obtained in a fine powder. This may be effected by slaking lime with just enough water to enable it to fall into a fine powder, which is then heated to redness in a crucible to drive off again the water. This dry powder is agitated frequently with the strongest rectified spirits for twenty-four hours. The alcohol is then distilled off, collecting the first portions. A repetition of this treatment will bring the alcohol to a high degree of strength. To deprive it of the last traces of water, it is necessary to add to it a sufficient quantity of freshly fused (cold) caustic potassa, and at once distil it over a naked fire until three-fourths of the liquid have passed over. It is now absolute alcohol.

Another and a simple method of strengthening alcohol, consists in exposing to the air a tightly closed bladder containing the spirits. The pores of the bladder permit the passage (and consequent evaporation) of a large portion of the water, but oppose that of the alcohol. Absolute alcohol is colorless, of a biting taste and pleasant smell. It has never been fairly frozen by the greatest artificial cold. It has a great affinity for water, which it greedily attracts from the air. This property makes it a violent poison, for when swallowed it deprives of water whatever it touches. Its affinity in this respect is shown by the heat evolved and by the contraction which takes place when alcohol and water are mixed. The greatest contraction is perceived when 53.7 measures of absolute alcohol are added to 49.8 measures of water. The sum, which is $103\frac{1}{2}$ measures, contracts to 100 measures. It is necessary to take this contraction into consideration when making alcoholic mixtures by measure. If we make them *by weight*, it of course differs not. The following experiment illustrates this contraction of alcohol and water: Take a long, thin tube, closed at one end, and carefully measure into it successively two fluid ounces of water, making a mark at each level. Pour out the last portion of two ounces, and add instead strong alcohol, (colored with turmeric, so that it may be seen,) to the upper mark. This addition must be made carefully, so that the alcohol does not mix but floats upon the water. Now close the tube, and invert it several times, until the two fluids be perfectly mingled, and it will be perceived that the mixture does not reach the level of the upper mark, demonstrating that a contraction has taken place.

Alcohol expands very much by heat and contracts by cold; which, added to its property of not freezing, makes it valuable for thermometers which are to be exposed to great cold. The dilatibility by heat, boiling point and specific gravity, afford three methods for determining the alcoholic strength of wines and spirits.

ALCOHOLIC STRENGTH OF SOLUTIONS.

1st. *By the dilatometer test.*—This test is founded upon the difference of dilatability of absolute alcohol and pure water. The small quantity of sugar and extraneous matters existing in the liquids under examination does not sensibly influence the correctness of the results. The dilatometer consists of a cylindrical glass vessel, resembling in appearance a glass hydrometer. The body of the vessel terminates below in a hair opening, which may be closed by a vulcanized pad and spring; above, it connects with a thermometer tube. The liquor to be tested is brought to the initial temperature for which the dilatometer has been graduated, say 77° Fah., and introduced into the instrument by plunging the latter, (with its lower aperture open,) therein. By suction the liquor is caused to fill the vessel, so that it rises in the thermometer tube to exactly the level of the line marked 0°. The lower aperture is then closed by releasing the spring, and the instrument is brought into a vessel of water of the higher temperature for which it is graduated, say 145°. The point to which the liquid rises in the thermometer stem will exhibit a number which will indicate the alcoholic strength of the liquor. These different per centage points are obtained by careful experiments performed upon alcoholic mixtures of different and accurately determined strength.

2d. *By the thermometer test.*—This method gives accurate results, which are not influenced materially by the sugar and salts of the liquors experimented upon. It requires a very delicate thermometer, which is graduated between the boiling points of alcohol and water. Tables have been constructed, or may be constructed by any one possessing such a thermometer, by taking the boiling point of alcoholic mixtures of known strength. By taking the boiling point of an unknown sample of liquor, its alcoholic strength is at once known upon reference to the table. Since the pressure of the atmosphere influences the boiling point of liquids, a table of corrections for the barometer must be used whenever the atmospheric pressure varies from 30 inches of mercury.

3d. *By the specific gravity test.*—This is the method in universal use. It is simple, accurate, speedy and not costly. The sugar, salts, &c., existing in fermented liquors influence its results and must be attended to. The specific gravity is taken either directly by comparing the weight of equal bulks of the liquid and water, or as is generally the custom, by means of the hydrometer, which gives the specific gravity by simple inspection. Since specific gravity enables us to test the strength of alcohol of saccharine solutions and of vinegar, it is important at this place to give a particular exhibition of the whole subject of specific gravity and hydrometers.

In ordinary language, the words "specific gravity" and "density," convey exactly the same idea, viz: the difference of weight of equal bulks of all bodies. Water is taken as the standard for this comparison, (for solids and liquids,) because it is universally accessible, and from the principle of Archimedes, to which allusion will be made directly. If, therefore, we weigh a cubic inch of every solid and liquid, and divide each result by the weight of the cubic inch of water, we will have a series of numbers, representing the specific gravity of the respective substances. In this series the specific gravity of water will be *one*, that of some bodies will be *less than one*, and that of the remainder *greater than one*. Thus:

ALCOHOLIC STRENGTH OF SOLUTIONS.

The specific gravity of water, . . . 1.000
" " lead, . . . 11.350
" " rolled platinum, . 22.070
" " marble, . . 2.840
" " oak wood, . . 1.170
" " cork, . . . 0.240
" " sulphuric acid, . . 1.848
" " ether, . . . 0.715

The most accurate method of taking specific gravities, is by actual weighing, which any one may readily perform with a "specific gravity bottle," and a sufficiently delicate balance. The following is a convenient way of making such a bottle where the regular one cannot be obtained. Take a glass stoppered bottle, of say two fluid ounces capacity, and file a longitudinal groove in the stopper. When the bottle is filled to the brim with liquid and this stopper gently dropped in, the excess of fluid will escape through the channel of the groove, and the stopper may be firmly set. One object to be gained is accomplished, this bottle will always measure exactly the same bulk of any liquid. By the aid of such a bottle, an apothecary's balance of ordinary delicacy, and accurate weights, any one can, by strictly following the directions which I shall give, determine the percentage of sugar, alcohol and acetic acid, in their respective aqueous solutions.

In weighing with such a balance, care should be taken to first test its accuracy, by bringing it into equilibrium with weights in each scale pan.

When these weights are shifted to opposite scale pans, the balance should be still in equilibrium; if it is not, the arms of the beam are of unequal length. We can, however, still use it, provided it be sensible to a small weight, by recourse to the system of double weighing. By this method, we place the bottle to be weighed in, let us say, the right hand scale pan, and counterbalance it exactly with shot and pieces of paper; then remove the bottle and add in its stead enough *weights* to bring the balance to equilibrium. The weights necessary will be the exact weight of the bottle.

In taking specific gravities by the bottle, it is imperative to have, once for all, the accurate weight of the dry empty bottle,* and that of the bottle full of pure water, (rain or distilled,) at the temperature of 39° Fah. This temperature of water is taken to determine specific gravities, because at it water possesses its greatest density. Below 39° Fah., water expands again, becomes less dense, until it is converted into ice. If this be not already known, a little reflection will show that it must be so, for otherwise ice would not float upon water. We can readily by means of ice and a thermometer, bring the water to be weighed in the specific gravity bottle to the temperature of 39°. Now, whenever we need to

---

* If wet the bottle may be dried by warming it, and sucking the air from it by a glass tube or a straw.

determine the specific gravity of any liquid, we have only to weigh our specific gravity bottle full of it. Then,

A. The difference of weight between the bottle full of water and empty, is the weight of one measure of water.

B. The difference of weight between the bottle full of the liquid under examination and the empty bottle, is the weight of one measure of said liquid.

Hence B divided by A, ($i.$ $e.$ $\frac{B}{A}$,) equals the specific gravity of the liquid in question.

In performing this process, the temperature of the liquid to be examined is brought to the temperature of 60° Fah., that being an ordinary temperature. In the different specific gravity tables which I shall give, the temperature to which the liquid must be brought to agree with them will be stated at the head of the table. If not so stated, 60° Fah. is understood.

I shall now give the method of obtaining the density of solids by the specific gravity bottle, because the principle upon which hydrometers depend, viz: "the principle of Archimedes," is directly concerned in it. At the risk of being considered trite, let me repeat the circumstances which, (it is said,) gave rise to the discovery of this principle. Archimedes lived B. C. 285–212. King Hiero, of Syracuse, had given to a goldsmith gold for the manufacture of a new crown, and suspected upon the completion of the work, that some of the gold had been stolen and silver added to make up the deficiency of weight. The matter was placed for investigation in the hands of Archimedes, who long pondered in vain over the solution of the problem. At length, one day upon stepping into a full bath, the water overflowed, which conveyed so forcibly to his mind a method of arriving at the desired information, that, in the most unphilosophical manner, naked as he was, (so stated, probably, to give point to the story,) he rushed home, crying, "Eureka, I have found it." The principle he discovered, and upon which hydrometers and the specific gravity method for solids depend, is, that "when a body is immersed in a fluid, the weight of the fluid displaced is equal to the weight which the body loses by the immersion." It enables us to obtain the weight of a quantity of water equal in bulk to the solid, which is all we want for arriving at the specific gravity. In other words, if we weigh a solid suspended by a hair, first in air, then in water, we will have the loss of weight in water, which is the weight of a quantity of water equal in bulk to the solid. Consequently, the weight in air divided by the loss of weight in water, will give a number which represents the specific gravity of the solid.

We can obtain the same result by the specific gravity bottle, in which case, we weigh the *water displaced*. Take the example of emery. Weigh some in fragments or powder in the dry bottle.

Now, without removing the emery, fill the bottle with pure water of the temperature of 39°. We know then the weight of the bottle full of water; the weight of the emery; and can get, by a simple calculation, the weight of pure water displaced by the emery; which, by the Archimedean principle, has a bulk equal to that of the emery; the quotient of this divided into the weight of the emery, gives its specific gravity. An example will render this plain:

Suppose the bottle holds of water, . . . 1000 grains.
The emery introduced, ... 100 "

Weight of whole, *had no water been displaced*, . 1100 "
But the observed weight is only . . . 1070 "

Hence the water displaced by the emery weighs, 30 "
and $\frac{100}{30}$ = 3·333 specific gravity of the emery.

Archimedes determined the specific gravity of gold and silver, 19.5 and 10.5 respectively; and next that of the crown. The latter, if no silver had been added, must be 19.5; if the suspected fraud had taken place, it must, of necessity, be less than 19.5; which was found to be the case, and thus, as in thousands of other instances, science was enabled to triumph over crime.

HYDROMETERS,

called also areometers, are instruments employed for determining the specific gravities of liquids. When prepared for use in particular liquids, they bear corresponding names, as the saccharometer, for solutions of sugar; the alcoholometer, for spirits; the acetometer, for vinegar, &c. Hydrometers must be made of a material unaffected by the liquid in which they are to be employed, as of glass for acids; silver, or hard rubber, for spirits, &c. The following is a representation of a glass hydrometer.

FIG. 1.

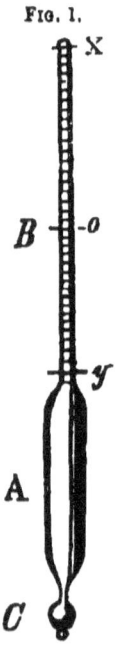

Like all the members of its class, it contains three parts: the body, A; the graduated stem, B; and the weight, C, which, by lowering its centre of gravity, enables it to float vertically in a liquid. In glass hydrometers, this weight is a bulbous extension of the body, and contains sufficient mercury, or small shot, to effect its purpose.

Let the instrument be so constructed that in pure water, of temperature 39° Fah., it sinks to 0. In sinking, it displaces water until, by the principle of Archimedes, a bulk of water equal in weight to the instrument is displaced. The weight of this bulk of displaced water and of the instrument hold each other in equilibrium; the hydrometer floats immersed to the point 0.

If we place the hydrometer in alcohol, which is specifically lighter than water, as its own weight and form are unchanged, it will sink deeper, or until the alcohol displaced equals the weight of the hydrometer, suppose to x. In vinegar, it will not have to sink as far as 0, but only let us say to y, for the specific gravity of that fluid is greater than water. The *different volumes* of alcohol displaced by the hydrometer sinking to x; of water when it sinks to o; of vinegar when it sinks to y; all weigh the same, viz: as much as the hydrometer weighs in air. It will readily be perceived, that we may graduate the stem so that the point to which it will sink will indicate the specific gravity of a fluid in which it is immersed. In the method of determining the specific gravities of fluids by the bottle, we compare different weights of the *same volume*. By the hydrometer, we compare the different *volumes* which the same weight of the respective fluids occupies, from which we infer their specific gravity, *i. e.* relative weights of the same volume. There are several methods of graduating the stem. In some, the degrees refer to a table, which gives the corresponding specific gravities; in others, the specific gravities are expressed directly upon the instrument; and, in others, the per centage of sugar, alcohol, acetic acid, &c., are marked upon the stem. The larger the body is in proportion to the stem, the more delicate is the hydrometer.

There are several methods by which the graduation points are obtained for hydrometers. The least philosophical of these is that of Beaumé, whose hydrometers have attained a universal use, especially in this country. They are used with tables, which give the specific gravity corresponding to every degree.

Beaumé's hydrometers are of two classes, one called "*pese-acides*," "*pese-sels*," &c., for liquids heavier than water; the other, "*pese-ether*," for liquids lighter than water. This obviates the necessity for instruments with a long delicate, and, consequently, fragile stem. The hydrometers of the first class, have the degrees numbered from below upward; in those of the second class, they run downward. It is useful at times to employ instruments of which the stem contains only a portion of either of these scales. This enables the stem to be more slender in proportion to the body, which gives a more sensitive instrument, as the degrees are thereby farther apart. The following are Beaumé's tables complete:

HYDROMETERS. 141

BEAUME'S HYDROMETER TABLE FOR LIQUIDS HEAVIER THAN WATER.

Temperature of liquid 54° *Fah.*

0	1·000	38	1·359
1	1·007	39	1·372
2	1·014	40	1·384
3	1·022	41	1·398
4	1·029	42	1·412
5	1·036	43	1·426
6	1·044	44	1·440
7	1·052	45	1·454
8	1·060	46	1·470
9	1·067	47	1·485
10	1·075	48	1·501
11	1·083	49	1·516
12	1·091	50	1·532
13	1·100	51	1·549
14	1·106	52	1·566
15	1·116	53	1·583
16	1·125	54	1·601
17	1·134	55	1·618
18	1·143	56	1·637
19	1·152	57	1·656
20	1·161	58	1·676
21	1·171	59	1·695
22	1·180	60	1·714
23	1·1 0	61	1·736
24	1·199	62	1·758
25	1·210	63	1·779
26	1·221	64	1·801
27	1·231	65	1·823
28	1·242	66	1·847
29	1·252	67	1·872
30	1·261	68	1·897
31	1·275	69	1·921
32	1·286	70	1·946
33	1·298	71	1·974
34	1·309	72	2·002
35	1·321	73	2·031
36	1·334	74	2·059
37	1·346	75	2·087

VINEGAR MANUFACTURE.

BEAUME'S HYDROMETER TABLE FOR LIQUIDS LIGHTER THAN WATER.

Temperature of liquid 54° Fah.

60	0·744	34	0·856
59	0·748	33	0·862
58	0·752	32	0·867
57	0·756	31	0·872
56	0·760	30	0·877
55	0·764	29	0·882
54	0·768	28	0·888
53	0·772	27	0·893
52	0·776	26	0·899
51	0·780	25	0·906
50	0·784	24	0·911
49	0·788	23	0·917
48	0·792	22	0·923
47	0·795	21	0·929
46	0·799	20	0·935
45	0·803	19	0·941
44	0·806	18	0·947
43	0·812	17	0·953
42	0·817	16	0·959
41	0·821	15	0·966
40	0·826	14	0·973
39	0·831	13	0·979
38	0·836	12	0·986
37	0·841	11	0·993
36	0·846	10	1·000
35	0·851		

The following is a good rule for converting the degrees of Beaumé for liquids denser than water into specific gravities, when the table is not at hand:

Let $B =$ the ascertained degree of Beaumé; then specific gravity of liquid $= \frac{144}{144-B}$

For example, suppose the liquid indicates 66° B., then $\frac{144}{144-66} = \frac{144}{78} = 1·846$ its specific gravity.

Gay Lussac's volumeter, is a hydrometer with a rational graduation. In it, the additional volume of liquid displaced by the instrument when it sinks one degree, bears a known ratio to the volume displaced by the instrument in water. This hydrometer deserves to be employed more generally than it is in our country. It is made in series of several instruments suited to liquids of different densities, thereby avoiding the frangibility of a single instrument with a long and delicate stem. The specific gravity is found by dividing one hundred by the number of degrees to which the volumeter sinks in the liquid under examination.

For example, in a certain liquid, heavier than water, it sinks to 80°; then specific gravity $= \frac{100}{80} = 1\cdot25$.

In another liquid, less dense than water, it sinks to 116; then, specific gravity $= \frac{100}{116} = 0\cdot862$.

Thus may a table be prepared to use with the volumeter, which will avoid even this simple calculation. This instrument, if correct, sinks in water of 39° Fah. to 100° of its scale.

The hydrometers in the market are often very false. In buying one, it should always be tested in some suitable fluid, of which the density has been accurately determined by the specific gravity bottle. Beaumé's hydrometer for *dense* fluids should sink in water to 0°, and that for *lighter* fluids should float at 10°.

The saccharometer is a hydrometer of which the

degrees indicate the per centage of sugar existing in aqueous solutions of this substance. Thus, 14° saccharometer indicates that the solution contains 14 per cent. by weight of sugar. On page 121 I have given a table in which the saccharometer degrees are translated into specific gravities.

The acetometer is a hydrometer for determining the per centage by weight of acetic acid in vinegar. The instrument has two scales, one indicating the per centage of anhydrous acetic acid, the other that of hydrated acetic acid. Specific gravity is a poor test for vinegar. A chemical acetometer will be described in the chapter on acetic acid.

The alcoholometer is a hydrometer which indicates the per centage of absolute alcohol *by volume* in spirits. I shall give tables, by means of which, the per centage by volume may be translated into per centage by weight, and vice versa. The liquor merchant buys and sells by measure; the vinegar maker calculates the per centage in acid of his manufactured article by weight, and hence it is important to learn in alcohol both kinds of per centage. Liquor dealers employ alcoholometers graduated differently from that just described. In them, the point of departure is not water, but "*proof spirits,*" that is, a certain definite mixture of alcohol and water, established by law, and differing in different countries. A stronger spirit is of so many degrees over proof, and a weaker one, of so many degrees under proof. English proof spirit has at 60° Fah. a density of 0·9186, containing a per centage of absolute alcohol = 49·50 by weight, and 57·27 by measure. The proof spirit of New York and of many other States, has at 60° Fah. a density of 0.9335, and contains in one hundred *measures* fifty of absolute alcohol and fifty of water. The proof spirit of Ohio has at 60° Fah., specific gravity, 0·9367 = 49 per cent., by volume, of absolute alcohol. The whole system of proof is cumbersome, and descends from an age, when the king's thumb measured the inch, and the weight of his fist the pound. In old times, spirits were poured upon gunpowder and set on fire; if the powder ignited, they were over-proof, and if not, they were called under-proof. The vinegar maker should be entirely ignorant of proof, and should test the spirits which he purchases with his own alcoholometer, which may be that either of Tralles or of Gay Lussac. Remember that the expression 30° Tralles, or 30° Gay Lussac, indicate that in one hundred gallons of the said spirits at 60° Fah., thirty gallons of absolute alcohol are contained.

In applying hydrometers, the liquid is placed in a tall cylindrical glass, and the hydrometer lowered gradually into it without wetting that portion of the stem rising above the liquid, which, by *increasing the weight* of the instrument, would give false results. The hydrometer must not adhere to the side of the vessel.

The ordinary style of alcoholometers will not do for testing weak solutions of alcohol. The lower degrees are too close together. For this purpose, a more delicate instrument should be employed, which, with a stem several inches long, contains only from 0°–12°. I will now give the tables of corrections for temperature, to enable a speedy and certain alcoholometer test. If the temperature of the spirits be exactly 12°·5 Réaumer = 60°·1 Fah., a simple inspection of the Tralles or Gay Lussac alcoholometer immersed therein, gives its per centage, by volume, of absolute alcohol. If the temperature varies from this point, the temperature of the spirits must be very carefully noted, and the proper correction of the table applied. Réaumer's thermometer is not employed in our country, but its degrees are given in the table because a very excellent Berlin Tralles instrument is made, of which the counterpoise or weight is the bulb of a Réaumer thermometer.*

* Bullock and Crenshaw, or McAllister & Bro., of Philadelphia, can furnish all of the instruments described in this work. A good Tralles instrument with Fahrenheit thermometer as described, costs $3.00.

HYDROMETERS. 147

ALCOHOLOMETER TABLE, CORRECTIONS FOR TEMPERATURES.

For temperatures below $12°\cdot5$ *Réaumer* $= 60°\cdot1$ *Fah.*

Degrees of Tralles read.	Nos. of degrees for which 1 per cent. of alcohol must be *added*.		Degrees of Tralles read.	Nos. of degrees for which 1 per cent. of alcohol must be *added*.	
	Réaumer.	Fah.		Réaumer.	Fah.
40	2·0	4·5	71	2·6	5·8
41	2·1	4·7	72	—	—
42	—	—	73	—	—
43	—	—	74	2·7	6·1
44	—	—	75	—	—
45	2·2	4·9			
			76	2·7	6·1
46	2·2	4·9	77	—	—
47	—	—	78	2·8	6·3
48	—	—	79	—	—
49	—	—	80	—	—
50	—	—			
			81	2·9	6·5
51	2·3	5·2	82	—	—
52	—	—	83	3·0	6·7
53	—	—	84	—	—
54	—	—	85	—	—
55	—	—			
			86	3·0	6·7
56	2·3	5·2	87	3·1	7·0
57	2·4	5.4	88	3·2	7·2
58	—	—	89	3·3	7·4
59	—	—	90	3·4	7·6
60	—	—			
			91	3·5	7·8
61	2·4	5·4	92	3·6	8·1
62	—	—	93	3·7	8·3
63	2·5	5·6	94	3·9	8·7
64	—	—	95	4·0	9·0
65	—	—	96	4·2	9·4
			97	4·5	10.1
66	2·5	5·6			
67	—	—			
68	2·6	5·8			
69	—	—			
70	—	—			

VINEGAR MANUFACTURE.

ALCOHOLOMETER TABLE, CORRECTIONS FOR TEMPERATURE.

For temperatures above $12°·5$ *Réaumer* $= 60°·1$ *Fah.*

Tralles degrees read.	Number of degrees for which 1 per cent. of alcohol must be subtracted.		Tralles degrees read.	Number of degrees for which 1 per cent. of alcohol must be subtracted.	
	Réaumer.	Fah.		Réaumer.	Fah.
40	2·0	4·5	71	2·5	5·6
			72	—	—
41	2·0	4·5	73	—	—
42	—	—	74	—	—
43	—	—	75	2.6	5·8
44	—	—			
45	—	—	76	2·6	5·8
			77	—	—
46	2·0	4·5	78	—	—
47	2·1	4·7	79	2·7	6·1
48	—	—	80	—	—
49	—	—			
50	—	—	81	2·7	6·1
			82	—	—
51	2·1	4·7	83	2·8	6·3
52	—	—	84	—	—
53	2.2	4·9	85	—	—
54	—	—			
55	—	—	86	2·9	6·5
			87	—	—
56	2·3	5·2	88	—	—
57	—	—	89	3·0	6·7
58	—	—	90	3·1	7·0
59	—	—			
60	—	—	91	3·1	7·0
			92	3·3	7·4
61	2·3	5·2	93	—	—
62	—	—	94	3·4	7·6
63	—	—	95	—	—
64	—	—			
65	—	—	96	3·6	8·1
			97	—	—
66	2·4	5·4	98	3·7	8·3
67	—	—	99	4·2	9·4
68	—	—	100	4·4	9·9
69	2.5	5·6			
70	—	—			

In these tables, the hyphens refer to the numbers immediately preceding, to avoid repetition and to catch the eye; thus, in the second table, for 72 — —, read 72°—2·5 Réaumer,—5·6 Fah., &c.

Let me now illustrate, by example, the employment of the tables. Suppose that in winter we test a spirit at the temperature of 8° Réaumer, and find its strength 83°, *i. e.* 83 per cent. of absolute alcohol. This per centage is too low, because the temperature is below the normal point of 12°·5, the spirits are denser at 8° than at 12°·5. The difference in temperature is 12°·5—8 = 4°·5. Now, by the table, for 83°, we must add 1 per cent. of alcohol for every 3° of Réaumer below 12°·5, hence for 4°·5 Réaumer, we must add $\frac{4\cdot5}{3°} = 1°\cdot5$. Hence the correct strength of the alcohol is 83°+1°·5 = 84°·5 per cent. of absolute alcohol.

Here is the same example in degrees of Fahrenheit. The temperature = 50° Fah., the indication 83 per cent. Fifty degrees is 10°·1 below 60°·1. If for 6°·7 Fah., we add 1 per cent. alcohol; for 10°·1, we must add $\frac{10\cdot1}{6\cdot7} = 1°\cdot5$ Hence 83 per cent. + 1°·5 = 84·5 as before. If we were selling this sample of spirits by the alcoholometer, without making the correction for temperature, we would lose one and a half gallons of absolute alcohol on every one hundred gallons of the spirits.

Let us take another example, where the tem-

150   VINEGAR MANUFACTURE.

perature is above the normal one, say 20° Réaumer, or 77° Fah., and let the alcoholic indication, (first column,) be 89°. In this case, we must, according to the table, diminish 89°, the observed per centage; thus, 20° Réaumer, — 12°·5 Réaumer, = 7°·5. By table, we must subtract 1 per cent. absolute alcohol for every 3° above 12°·5, and, hence, for $7°·5 \frac{7·5}{30} = 2°·5$. 2°·5 Tralles, are therefore to be subtracted from 89°, which leaves 86°·5 as the correct per centage of alcohol in the spirits.

With Fahrenheit degrees, the same example would be thus worked out:

$77° — 60°·1 = 16°·9; \frac{16·9}{6·7} = 2°·5; 89° — 2°·5 = 86°·5.$

The following useful table enables us to obtain the per centage strength of alcoholic solutions, by determining the specific gravity either with the hydrometer, or by the specific gravity bottle:

SPECIFIC GRAVITY TABLE.

Of alcoholic mixtures (with water) by weight and volume, at temperature 60°·1 Fah.

Per centage of absolute alcohol.	Specific gravity for		Per centage of absolute alcohol.	Specific gravity for	
	Volume per cent.	Weight per cent.		Volume per cent.	Weight per cent.
1	0·9985	0·9981	9	0·9878	0·9855
2	0·9970	0·9965	10	0·9866	0·9841
3	0·9956	0·9947	11	0·9854	0·9828
4	0·9942	0·9930	12	0·9843	0·9815
5	0·9928	0·9913	13	0·9832	0·9802
6	0·9915	0·9898	14	0·9821	0·9789
7	0·9902	0·9884	15	0·9811	0·9778
8	0.9890	0·9869	16	0·9800	0·9766

HYDROMETERS. 151

SPECIFIC GRAVITY TABLE, CONTINUED.

Percentage of absolute alcohol.	Specific gravity for		Percentage of absolute alcohol.	Specific gravity for	
	Volume per cent.	Weight per cent.		Volume per cent.	Weight per cent.
17	0·9790	0·9753	59	0·9156	0·8979
18	0·9780	0·9741	60	0·9134	0·8956
19	0·9770	0·9728	61	0·9112	0·8932
20	0·9760	0·9716	62	0·9090	0·8908
21	0·9750	0·9704	63	0·9067	0·8886
22	0·9740	0·9691	64	0·9044	0·8863
23	0·9729	0·9678	65	0·9021	0·8840
24	0·9719	0·9665	66	0·8997	0·8816
25	0·9709	0·9652	67	0·8973	0·8793
26	0·9698	0·9638	68	0·8949	0·8769
27	0·9688	0·9623	69	0·8925	0·8745
28	0·9677	0·9609	70	0·8900	0·8721
29	0·9666	0·9593	71	0·8875	0·8696
30	0·6655	0·9578	72	0·8850	0·8672
31	0·9643	0·9560	73	0·8825	0·8649
32	0·9631	0·9544	74	0·8799	0·8625
33	0·9618	0·9528	75	0·8773	0·8603
34	0·9605	0·9511	76	0·8747	0·8581
35	0·9592	0·9490	77	0·8720	0·8557
36	0·9579	0·9470	78	0·8693	0·8533
37	0·9565	0·9452	79	0·8665	0·8508
38	0·9550	0·9434	80	0·8639	0·8483
39	0·9535	0·9416	81	0·8611	0·8459
40	0·9519	0·9396	82	0·8583	0·8434
41	0·9503	0·9376	83	0·8555	0·8408
42	0·9487	0·9356	84	0·8526	0·8382
43	0·9470	0·9335	85	0·8496	0·8357
44	0·9452	0·9314	86	0·8466	0·8331
45	0·9435	0·9292	87	0·8436	0·8305
46	0·9417	0·9270	88	0·8405	0·8279
47	0·9399	0·9249	89	0·8373	0·8254
48	0·9381	0·9228	90	0·8339	0·8228
49	0·9362	0·9206	91	0·8306	0·8199
50	0·9343	0·9184	92	0·8272	0·8172
51	0·9323	0·9160	93	0·8237	0·8145
52	0·9303	0·9135	94	0·8201	0·8118
53	0·9283	0·9113	95	0·8164	0·8089
54	0·9263	0·9090	96	0·8125	0·8061
55	0·9242	0·9069	97	0·8084	0·8031
56	0·9221	0·9047	98	0·8041	0·8001
57	0·9200	0·9025	99	0·7995	0·7969
58	0·9178	0·9001	100	0·7946	0·7938

The specific gravity column of the volume per cents. is by Brix, calculated according to Tralles, where the specific gravity of water is taken equal to 1 at 60°·1 Fah., and whence that of absolute alcohol is 0·7946. The column of weight per cents. is by Fownes. In it the specific gravity of water is taken $= 1$ at the proper temperature 39° Fah., and becomes 0·9991 at 60° Fah., whence the specific gravity of absolute alcohol on that basis is at 60°·1 $= 0·7938$. This accounts for the very small discrepancies between the two columns.

To illustrate the use of this table, suppose we talk of an alcohol of 46 per cent.; if we mean per centage by volume, its density $= 0·9417$; if we understand per centage by weight, its specific gravity would be 0·9270.

Again: suppose that we have determined the specific gravity of an alcohol to be 0·8825; reference to the table will show that its per centage strength of absolute alcohol is 73 by *volume*, and between 65 and 66 by *weight*.

Finally: suppose, knowing that a spirit contains 44 per cent. absolute alcohol by volume, we desire to learn what per centage by weight this would be equivalent to. Opposite 44 in the first column we find in the second column 0·9452. This number in the third column corresponds to a per centage of 37, which is a weight per centage. Weight per cents. are converted into volume per cents. by a similar manner.

There are rules for mutually transforming

weight and volume per cents. from the table with great accuracy, thus:

Rule.—To change volume per cent. to weight per cent. Multiply the volume per cent. (*i. e.* degrees of Tralles,) by 0·794, *i. e.* the specific gravity of absolute alcohol, and divide by the specific gravity which corresponds to the volume per centage.

Example.—For 82 volumes per cent., how many weights per cent?

$$\frac{82 \times 0.794}{0.8583} = 75.85 \text{ per centage by weight.}$$

Rule.—To transform *weight* per cents. to *volume* per cents. Multiply the weight per cent. by its corresponding specific gravity and divide by 0·794, *i. e.* by the specific gravity of absolute alcohol.

Example.—What per centage by volume does 76 per centage by weight correspond to?

$$\text{Answer. } \frac{76 \times 0.8581}{0.794} = 82.1$$

In the following short tables this calculation is performed for the lower per cents. of spirits of the strength employed by the vinegar maker.

TABLE I.

Vol. per ct.	=	Weight per ct.
1	=	0·80
2	=	1·60
3	=	2·40
4	=	3·20
5	=	4·00
6	=	4·80
7	=	5·60
8	=	6.40
9	=	7·24
10	=	8·03
11	=	8·88
12	=	9·70

TABLE II.

Weight per ct.	=	Vol. per ct.
1	=	1·25
2	=	2·50
3	=	3·75
4	=	5·00
5	=	6·25
6	=	7·50
7	=	8·70
8	=	9·56
9	=	11·20
10	=	12·40

For example 6 per cent. by volume corresponds to 4·8 per cent. by weight. 5 per cent. by weight corresponds to 6·25 per cent. by volume.

Some alcoholometers have upon the same stem two scales, that of Tralles, which indicates per centage by volume, and that of Richter, which tells weight per cents. The latter scale is worthless, being constructed upon a false principle.

THE CALCULATION FOR MAKING DEFINITE MIXTURES OF ALCOHOL AND WATER.

The manufacturer of vinegar who employs spirits, buys them of a certain strength, paying according to the quantity of absolute alcohol which they contain. He dilutes them with water to a definite point, according to the acid strength which the vinegar must have. He must, by all means, know how much water to add in any given case.

Alcohol and water contract when mixed. The liquor dealer takes this contraction in account, but not so the vinegar maker, because the error involved is trifling on account of the weakness of his alcoholic solutions.

Suppose it is required to obtain one hundred gallons of spirits of 5%* Tralles by diluting alcohol of 80% Tralles. How much alcohol and how much water must be taken?

Rule.—Multiply the number of gallons of the required mixture by its required alcohol per cent-

---

* The sign % indicates per centage.

age and divide by the alcohol per centage of the spirit used. The quotient gives the number of gallons of the alcohol which must be taken for the mixture.

Thus, in the above case, $\frac{100 \times 5}{80} = 6\cdot 25$. Then take 6·25 gallons of 80% alcohol, and add 100—6·25 = 93·75 gallons of water to it; or add water to the above quantity of spirits until we have a hundred gallons of the mixture. There is another way of obtaining the same result:

Rule.—Divide the per centage (Tralles) of the strong spirits by the required per centage of the mixture. The quotient will express the number of gallons of mixture which *one gallon* of the strong spirits will afford. Thus, in the last example, $\frac{80}{5} = 16$. That is one gallon of 80% spirits will make sixteen gallons of 5% mixture. Ten gallons of the former will make one hundred and sixty of the latter, or in general terms:

$$16 : 1 :: a : x = \tfrac{a}{16}$$

where x denotes how much spirits must be taken, and a represents the required number of gallons of the mixture. In the former example, one hundred gallons of mixture were desired, then $x = \frac{100}{16} = 6\cdot 25$. Hence 6·25 gallons of 80% spirits must be mixed with enough water to make one hundred gallons, as obtained by the former rule.

Suppose thirty-eight gallons of the mixture were wanted, then $\frac{38}{16} = 2\frac{37}{100}$. Add water to $2\frac{37}{100}$ of 80% spirits to make the mixture measure 38 gallons.

ALCOHOL TEST FOR WINES AND BEER.

These beverages contain sugar, gum, mineral salts, &c., which affect the specific gravity of the liquid; consequently the alcoholometer will not immediately indicate their alcoholic strength. The desired information may be obtained in two ways.

I. Three hundred measures (any convenient measure) are distilled in a little still of tinned copper, heated by a spirit lamp. This still is made for the purpose, and its worm must be perfectly cooled, so as not to lose any alcohol. The distilled liquid is caught in a vessel graduated on the same scale as that by which the beverage was measured. As soon as 100 measures of spirits have passed over, they are tested with the alcoholometer, and $\frac{1}{3}$ its per centage found indicates the alcoholic strength of the beverage under examination. If the liquid tested is poor in alcohol, distill off only 50 measures, and take $\frac{1}{6}$ its alcoholometer indications. If it is very rich, distill off $\frac{1}{2}$ or $\frac{2}{3}$, and take $\frac{1}{2}$ or $\frac{2}{3}$ the per centage given by the alcoholometer.

II. METHOD.—This method is not absolutely accurate; but sufficiently so for practical purposes. Filter a portion of the beverage, and determine its specific gravity very accurately at 60° Fah. by the sp. gr. bottle. Now weigh out more than enough of the filtered liquid to fill the sp. gr. bottle; boil off all the alcohol, and add pure water to restore exactly its original weight. Determine the specific gravity of this result also by the bottle. Subtract from 1 the *difference* of these two specific gravities, the remainder will be a specific gravity from which the alcoholic strength of the beverage may be inferred by using the Table on page 150.

For example, suppose the fermented liquid have a specific gravity before boiling $= 1\cdot0016$; after boiling $1\cdot0088$.

The difference of these numbers$=0\cdot0072$; 1 less, $0\cdot0072=0\cdot9928$.

Now by the Table on page 150, $0\cdot9928$ is nearest to $0\cdot9930$ in column 3, which corresponds to 4 per cent. by weight of absolute alcohol. In column 2d, it corresponds to 5 per centage by volume.

THE SACCHAROMETER TEST FOR ALCOHOL.

There is a method of ascertaining the alcoholic value of fermented liquids by the use of the saccharometer during the fermenting process. I have postponed the consideration of this subject to the present place where it may be more conveniently set forth.

The subject is called the "*attenuation of worts.*"

The saccharometer placed in beer worts (of the temperature for which the instrument is graduated,) immediately after the addition of yeast, will float at so many degrees. In solutions of pure sugar, these degree would indicate the per centage by weight of sugar in the solution, but not so in beer wort, which contains, besides sugar, gum, diastase, mineral salts, &c. In proportion as the fermentation proceeds, whether in sugar solutions or in beer worts, the specific gravity falls, because the denser sugar disappears, and the specifically lighter alcohol takes its place. Consequently, after fermentation, the saccharometer will indicate a lower per centage. This diminution of density is called the "*apparent attenuation of the worts.*" It does not indicate absolutely how much alcohol is formed, but how much more in one case than in another. For instance, suppose that a wort indicating 12° saccharometer is fermented in two portions and that after fermentation, one portion indicates 1° by the saccharometer, the other portion 3°. The "*apparent attenuation*" of the first portion would be 12°—1°=11°, and of the second portion 12°—3°=9°.

In other words, the greater apparent attenuation would indicate a more perfect fermentation.

To arrive at a knowledge of the real alcoholic strength of the fermented liquor, we must learn the "*real attenuation of the worts,*" *i. e.* how much sugar has really disappeared in the fermentation. We can then calculate the alcohol formed, for we know that a pound of sugar will yield a half a pound of absolute alcohol.

The real attenuation is readily ascertained by the following simple experiment. Take the saccharometer indication before fermentation. After fermentation, boil down to about one-half an accurately weighed portion of the clear malt wine, taking pains that none is lost by spirting; then restore to the exact weight taken by the addition of water. Test this at the proper temperature, by the saccharometer. In the latter fluid nothing is gone from the original worts but the alcohol. The last saccharometer result is due to the gum, salts, and other heavy substances in the malt wine, which were of course also present in the unfermented wort. Consequently the difference between the two saccharometer indications gives the per centage of sugar that has disappeared during fermentation. For example, suppose the worts indicated 12° saccharometer, the boiled malt wine 2°·4, then 12°—2°·4=9°·6 is the real attenuation. In other words, 9°·6 per cent. of sugar have been converted into absolute alcohol. Since sugar yields one-half its weight of alcohol, the fermented liquid contains $\frac{9°·6}{2} = 4°·8$ per cent. *by weight* of alcohol. By the Table on page 153, 4°·8 per cent. by weight corresponds to 6 per cent. by volume. The resulting beer then contains 6 per cent. by volume of absolute alcohol. This rule, though simple and sufficiently correct for practice, is not absolutely so; for it leaves out of calculation the specific gravity of the new yeast formed during the fermentation. Balling has prepared the following table of factors which obviates the error.

REAL ATTENUATION TABLE.

Saccharometer strength of the worts before fermentation.	Alcohol factors for "REAL ATTENUATION."	
	For solutions of crystalline sugar.	For solutions of beer worts.
6	0·5265	0·4993
7	0·5292	0·5020
8	0·5319	0·5047
9	0·5346	0·5074
10	0·5373	0·5102
11	0·5401	0·5130
12	0·5429	0·5158
13	0·5457	0·5187
14	0·5486	0·5215
15	0·5515	0·5245
16	0·5545	0·5274

To use this table, take the former example, in which the "*real attenuation*" was 9°·6. Now, instead of dividing by two, multiply by the factor opposite the first saccharometer indication, which was 12° in our example. Multiply the real attenuation then, since it is beer worts, by 0·5158, and we have 9°·6 × 0·5158 = 4°·95, the alcoholic percentage (by weight) of the malt wine. It will be seen that this result does not differ very much from 4°·8 as obtained by the division of the real attenuation by 2, as per last rule. Had the wort been made with *pure sugar*, we must multiply by a factor from the second column, viz.: (in the same example) by 0·5429. Here is another table by Balling, which gives the alcoholic strength from the "*apparent attenuation*," *i. e.* without resorting to the boiling experiment.

SACCHAROMETER TEST FOR ALCOHOL. 161

APPARENT ATTENUATION TABLE.

Saccharometer strength of the worts before fermentation.	Alcohol factors for "APPARENT ATTENUATION."	
	For solutions of crystalline sugar.	For solutions of beer worts.
6	0·4312	0·4073
7	0·4330	0·4091
8	0·4348	0·4110
9	0·4267	0·4129
10	0·4376	0·4141
11	0·4405	0·4167
12	0·4424	0·4187
13	0·4444	0·4206
14	0·4464	0·4226
15	0·4484	0·4246
16	0·4504	0·4267

Suppose the saccharometer indicates 12° and 1° respectively before and after fermentation, 12—1 = 11 is the apparent attenuation. Multiply this number (for beer worts) by the factor opposite 12 in the table, 0·4187×11=4·6 per centage by weight of absolute alcohol in the fermented liquid.

Suppose, again, the indication of the worts before fermentation being 12°, that the apparent attenuation were 9, then 0·4187×9=3·76 per cent. by weight of the malt wine.

CHAPTER IV.

ACETIC ACID.

ACETIC ACID is derived from alcohol by an oxidation process, which confines its action to the hydrogen, in consequence of which no separation of carbon takes place. This effect is brought about when weak alcohol, in contact with certain ferments, (of which vinegar is the best,) or in contact with spungy platinum, is subjected to the action of air at a slightly elevated temperature. The following is Liebig's theory of the vinegar process.

One atom of absolute alcohol is changed to one atom of hydrated acetic acid by the action of four atoms of oxygen of the air. Of these 4 oxygen atoms, 2 unite with 2 of the hydrogen of the alcohol, forming 2 atoms of water, and leaving a body called aldehyde, which unites with the two remaining oxygen atoms to form the acetic acid.

In the presence of plenty of oxygen, the aldehyde is never observed in the free state, as it is immediately transformed into acetic acid; but when there is too little air, its presence is quickly detected in the vinegar room by a penetrating aroma pervading the apartment, and by the eyes smarting.

The following symbols illustrate the vinegar process.

ACETIC ACID. 163

Alcohol . . . . $C_4H_6O_2$
Less H_2 (which $+O_2=2HO$ or water) . . H_2
Leaves *aldehyde*, which . . . . $C_4H_4O_2$
Plus oxygen . . . . . . O_2
Gives rise to hydrated acetic acid . . $C_4H_4O_4$

Here is the same process expressed numerically. From 100 pounds of absolute alcohol which consist of

Pounds Carbon, . . . 52·2
" Hydrogen, . . 13·0
" Oxygen, . . . 34·8
_____
100·0

4·3 pounds of hydrogen are withdrawn by 34·8 pounds of oxygen from the air, giving rise to 39·1 pounds of water, and leaving aldehyde, which consists of:

Pounds Carbon, . . . 52·2
" Hydrogen, . . 8·7
" Oxygen . . . 34·8
_____
95·7

With the aldehyde, 34·8 pounds of oxygen from the air unite, giving rise to hydrated acetic acid, which contains:

Pounds Carbon, . . . 52.2
" Hydrogen, . . 8·7
" Oxygen, . . . 69·6
_____
130·5

From this it is seen that 100 pounds of absolute alcohol give rise to 130½ pounds of hydrated acetic acid, that is the strongest possible acetic acid, which contains :

Anhydrous acetic acid, . . 85
Water, ... 15
_____
100

These 15 parts of water cannot be removed unless they are replaced by some base. In symbols

Hydrated Acetic Acid $= C_4H_4O_4$ -
= Anhydrous Acetic Acid, $C_4H_3O_3$
Plus water   HO
_____
$C_4H_3O_3 + HO$

In anhydrous acetate of soda, soda takes the place of the water of hydrated acetic acid. Its composition is ($C_4H_3O_3 + NaO$.)

130½ pounds of hydrated acetic acid contain 110·92 pounds of anhydrous acid, or in round numbers, 111 pounds; consequently we may say that every 100 pounds of absolute alcohol yield 111 pounds of anhydrous acetic acid.

One pound of *hydrated* acetic acid is generated from 0·77 pounds of absolute alcohol.

One pound of *anhydrous* acetic acid arises from 0·9 pounds of absolute alcohol.

These results are theoretical; in practice less strength is obtained for the vinegar, owing to a loss of alcohol by evaporation during the process.

In the quick vinegar process three particulars are to be observed.

1. *The nature of the ferment.*—Ready made vinegar is the best ferment; but sour beer, bread steeped in vinegar, leaven, &c., are also used. In these it is the nitrogenized or albuminous substance, which, with the vinegar, forms the ferment; chemically pure vinegar does not alone act as a ferment in the vinegar process.

2. *The strength of alcohol in the vinegar mixture.*

ACETIC ACID. 165

—In general terms the weaker the mixture (within certain limits) the more profitable is the manufacture. *It should not exceed* 10 *per cent. alcoholic strength.* In a strongly alcoholic mixture, not only is the loss by evaporation of the alcohol greater; but this body appears to neutralize the energy of the vinegar fermentation.

3. The limits of temperature for success are 72° Fah. and 100° Fah.

Within these limits the higher the temperature and the more air brought to the alcoholic solution, the quicker is its transformation to vinegar. This of course involves a loss of alcohol by evaporation, which loss is balanced by the speed of the process, enabling the capital invested to be the more frequently turned. The vinegar maker aims to convert his alcohol to vinegar as quickly as possible, with the least loss of alcohol by evaporation, and with the least remainder of unchanged alcohol in his product. The vinegars are weak solutions of acetic acid, flavored with certain aromatic ethers, which arise during the alcoholic and acetic fermentations. The strongest acetic acid crystalizes in flaky crystals, and is called "glacial acetic acid;" also, "radical vinegar." This is hydrated acetic acid with the least possible water. It boils at 248° Fah., volatilizing unchanged. It is colorless; has a peculiar sharp penetrating smell and burning taste. Its vapor may be kindled, giving a pale blue flame. It is heavier than water, its sp. gr. being at 59° Fah., 1·057 according to Van der Toorn, and 1·063 according to Mollerat. Its aqueous solutions have the peculiarity that the one containing between 67—69 per cent. of the anhydrous acid has the greatest specific gravity, as may be seen from the following table :

TABLE OF DENSITY

Of aqueous solutions of Acetic Acid at 59° *Fah.*

Anhydrous acetic acid percentage by weight.	Specific gravity.	Anhydrous acetic acid percentage by weight.	Specific gravity.	Anhydrous acetic acid percentage by weight.	Specific gravity.
1	1·0019	30	1·0485	58	1·0740
2	1·0037	31	1·0498	59	1·0745
3	1·0055	32	1·0510	60	1·0749
4	1·0072	33	1·0522	61	1·0753
5	1·0089	34	1·0537	62	1·0756
6	1·0107	35	1·0546	63	1·0759
7	1·0124	36	1·0558	64	1·0762
8	1·0141	37	1·0569	65	1·0764
9	1·0159	38	1·0580	66	1·0765
10	1·0177	39	1·0591	67	1·0766
11	1·0194	40	1·0601	68	1·0766
12	1·0211	41	1·0611	69	1·0766
13	1·0228	42	1·0621	70	1·0765
14	1·0245	43	1·0631	71	1·0763
15	1·0261	44	1·0640	72	1·0759
16	1·0277	45	1·0649	73	1·0754
17	1·0293	46	1·0658	74	1·0748
18	1·0310	47	1·0667	75	1·0741
19	1·0326	48	1·0675	76	1·0732
20	1·0342	49	1·0685	77	1·0722
21	1·0358	50	1·0691	78	1·0710
22	1·0373	51	1·0698	79	1·0696
23	1·0389	52	1·0705	80	1·0681
24	1·0404	53	1·0711	81	1·0664
25	1·0419	54	1·0717	82	1·0646
26	1·0433	55	1·0723	83	1·0626
27	1·0447	56	1·0729	84	1·0603
28	1·0460	57	1·0735	85	1·0574
29	1·0472				

ACETIC ACID. 167

Note that the solutions containing respectively 37 and 85 per centage of acetic acid have about the same density.

In testing stronger solutions of Acetic acid than those of 36 per cent., by the specific gravity, there will be a doubt as to the correct per centage value, as there may be two such values (see Table) to the ascertained specific gravity, In such cases we must ascertain whether a little water added to a portion *increases* or *diminishes* the found specific gravity. If it *increases* the same, then the vinegar is of the *higher* per centage.

The vinegar maker avoids this ambiguity, as he deals with solutions under 10 per cent. A hydrometer may be employed to ascertain the specific gravity, and consequently (by the foregoing table) the acid per centage of pure vinegar. Beaumé's instrument will not answer; it is not sufficiently delicate. A particular one, an *Acetometer*, must be procured, having the stem graduated between specific gravity 1 and 1·0177. The vinegar hydrometer is generally graduated to indicate at once the per centage of acetic acid; and it should be stated on the instrument whether the per cents. are of *hydrated* or of an hydrous acid. The temperature should also be stated on the instrument. It is generally 60° Fah. The vinegar before testing must be brought to this temperature, or else its temperature noted, and the indication corrected by the table of corrections, which generally accompanies the instrument.

The vinegar hydrometer, or specific gravity Acetometer, however, has a very restricted application, and is of much less use than the public generally believe.

It will not indicate the true strength of vinegars made from beer, wine, or cider, as these contain gum, sugar, mineral salts, &c., which increase the specific gravity. If such a vinegar contains 3 per cent. of acid, and 1 per cent. only of sugar, it will have a specific gravity which will give it an apparent strength of 6 per cent. acid by the Acetometer. The manufacturer, by the quick process, has a difficulty of an opposite nature to contend with, by reason of which the Acetometer sometimes makes his product to appear weaker than it really is. This vinegar sometimes contains unchanged alcohol, which lowers its specific gravity. I have seen vinegar hydrometers sink out of sight in vinegars of considerable acid strength. The specific gravity Acetometer must therefore be used, if at all, intelligently. The vinegar manufacturer who employs the quick process, and uses nothing but alcohol and water, may ascertain by the vinegar hydrometer whether much unchanged alcohol exists in his vinegar by determining its acid strength by one of the accurate methods about to be described, and then testing with the vinegar hydrometer, to see whether the corresponding specific gravity is indicated.

As the per centage of Acetic Acid in vinegar refers sometimes to *hydrated*, sometimes to *an-*

ACETIC ACID.

hydrous acid, it becomes necessary to know how to convert one of these values into the other. The following ratio between the two is sufficiently accurate for practical purposes.

Hydrated Acetic : Anhydrous Acetic :: 13 : 11.

Whence Hydrated Acetic $= \frac{\text{Anhydrous Acetic} \times 13}{11}$

and Anhydrous Acetic $= \frac{\text{Hydrated Acetic} \times 11}{13}$

Example.—How many per cents. of Anhydrous Acetic Acid do 6 per cent. of Hydrated Acid correspond to? *Answer,* $\frac{6 \times 11}{13} = 5\cdot1 =$ per cent. Anhydrous Acid. The following Tables save the trouble of this calculation.

TABLE I.

Hydrated acid per cent. to Anhydrous acid per cent.		Hydrated acid per cent. to Anhydrous acid per cent	
Hydrated acid per cent.	equal { Anhydrous acid per cent.	Hydrated acid per cent.	equal { Anhydrous acid per cent.
1	0·85	16	13·60
2	1·70	17	14·45
3	2·55	18	15·30
4	3·40	19	16·15
5	4·25	20	17·00
6	5·10	21	17·85
7	5·95	22	18·70
8	6·80	23	19·55
9	7·65	24	20·40
10	8·50	25	21·25
11	9·35	26	22·10
12	10·20	27	22·95
13	11·05	28	23·80
14	11·90	29	24·60
15	12·75	30	25·50

TABLE II.

Anhydrous acid per cent. to Hydrated acid per cent.		Anhydrous acid per cent. to Hydrated acid per cent.	
Anhydrous acid per cent. equal	Hydrated acid per cent.	Anhydrous acid per cent. equal	Hydrated acid per cent.
1	1·17	14	16·46
2	2·35	15	17·63
3	3·52	16	18·80
4	4·70	17	19·98
5	5·88	18	21·16
6	7·06	19	22·34
7	8·23	20	23·52
8	9·40	21	24·70
9	10·58	22	25·88
10	11·76	23	27·05
11	12·94	24	28·22
12	14·11	25	29·40
13	15·29	26	30·58

TESTS OF VINEGAR.

As vinegar is a substance entering so largely into the food of mankind, it becomes important to be able to ascertain whether it contains any deleterious matter through careless management or by fraudulent adulteration. The latter is unfortunately of frequent occurrence, since it is so easy to increase its *acidity* by the addition of some mineral acid. The following acids have been employed for the purpose.

Sulphuric Acid, (oil of vitriol.) Detected by boiling some of the vinegar in a porcelain or glass vessel, with a little solution of chloride of calcium. If this acid be present, a white precipitate of sulphate of lime will fall. Chloride of calcium solution is made by adding some water to pure hydrochloric acid, and then powdered limestone or chalk, until no effervescence is perceived by the further addition of chalk or limestone. The solution is filtered. Solutions of chloride of barium and nitrate of baryta, are also employed to test vinegar; but it must be borne in mind that they cause precipitates with certain sulphates which are always present in some vinegars. These sulphates exist in the water employed in the manufacture, or in the juices of the fruit from which the wine, cider, malt wine, &c., were made. By weighing a portion of the vinegar, adding a little free hydrochloric acid, and then the baryta salt, if any precipitate occur, we may filter it, and determine its weight by the process of analytical chemistry, to ascertain whether it is of unusual quantity.

Sulphurous acid may be detected by precipitating the sulphuric acid of the sulphates and free sulphuric of the vinegar by baryta water, (solution of hydrated oxyde of barium,) and then after filtering off these, by adding to the clear solution arsenic acid, which converts the sulphurous acid into sulphuric. The latter may then be determined by precipitation with chloride of barium.

Nitric acid is not often used to adulterate vinegar. Its presence may be detected by boiling the vinegar with a *little indigo*, or preferably with a few drops of indigo solution. The blue color is converted into a yellow one by nitric acid.

Hydrochloric acid, which is not often used for

adulteration, may be detected in the vinegar by the addition of nitrate of silver. If this acid be present, a white curdy precipitate, blackening in the light, falls. If the vinegar contains soluble chlorides, the same precipitate falls.

The presence of copper or lead in vinegar is perceived by passing through it a current of sulphuretted hydrogen gas, when a black precipitate takes place. Separate this precipitate by filtration, and dissolve it by heating with a few drops of pure nitric acid, then add a little water. To half of this add solution of yellow prussiate of potassa, which gives a brown precipitate when copper is present. To the rest add iodide of potassium, which yields a yellow precipitate with lead salts.

Certain acrid vegetable substances, such as grains of paradise, mustard seed, pepper, &c., are occasionally used for adulterating vinegar. Their biting acrid taste may be perceived by evaporating some of the vinegar to an extract.

ACETOMETRY.

The methods by which we are enabled to determine accurately the acetic acid strength of vinegar are the subjects of Acetometry. The specific gravity test, except that of Balling and others, which will be given towards the close of this chapter, is totally unworthy of use. The tests depending upon chemical principles, have been so improved within the last few years that they are

ACETOMETRY. 173

undoubtedly by far the most reliable, and they are so simple that any one possessing the requisite instruments may perform them with great facility, speed and accuracy.

I have thought proper, therefore, in the remainder of this chapter, to set forth, in detail, the several methods of chemical Acetometry,* and it will be unfortunate if any one interested, find not among them one suited to his purpose.

Chemical Acetometry depends upon the well-known fact that a given weight of the same alkali is always neutralized by a known weight of acetic acid. We are able by the dye litmus to tell when the point of neutralization has been reached. If then, to a certain weight of vinegar we add known weights of some alkali, until neutralization has been effected, we are able, from the weight of the alkali employed, to calculate the acetic acid strength of the vinegar.

There are several methods of performing this test. Those that enable us to arrive at the respective weights *by the aid of measures*, are in practice the most readily carried out.

The various alkalies employed are crystalized carbonate of soda; anhydrous, or dry carbonate of soda; carbonate of potassa and caustic ammonia.

By the different methods, the ingredients are either *all* weighed, or one only is weighed and the rest are measured. In weighing, an apothecaries'

*From Otto's excellent Lehrbuch der Essig Fabrikation.

prescription scales, and weights of good quality are required. In the following descriptions, the weights employed are the ounce, (apothecaries' weight,) which equals 480 grains, and the French gramme, 1 gramme = 15·44 grains.

I. METHOD BY DRY ALKALIES.

1. *By Crystalized Carbonate of Soda.*—These crystals contain carbonate of soda combined with water of crystalization. They must be glossy, clear, and all white powder (effloresced salt) should be scraped off. They must be rubbed into a coarse powder and placed in a wide-mouthed, glass stoppered bottle of 2 ounces capacity which is small and light enough to be weighed upon the balance employed. 100 grains of crystalized carbonate of soda will neutralize 35·7 grains of anhydrous acetic acid, and 42 grains of hydrated acetic acid. If, therefore, we weigh out 2 ounces, (*i. e.* 960 grains) of vinegar, every 27 grains of the soda will neutralize in it 1 per cent. of *anhydrous* acetic acid; and $22\frac{8}{10}$ (in round numbers 23) grains of the soda will neutralize 1 per cent. of *hydrated* acetic acid.

We have then only to observe the weight of carbonate of soda employed to neutralize the acid in two ounces of vinegar, and divide that weight by 27 to get the percentage of anhydrous acid; or by 23 to obtain the percentage of hydrated acid in the vinegar. For example, suppose that 2 ounces by weight of the vinegar require 135 grains of crystalized carbonate of soda for neutralization.

The vinegar then contains $\frac{135}{27}$ = 5 per cent. anhydrous acetic acid.
" " $\frac{135}{23}$ = 6 per cent. hydrated acetic acid.

The principle involved in this method is very simple; the main difficulty in practice is to ascertain the *exact point* of neutralization. To effect this purpose, solution of litmus and litmus papers are employed. Commercial litmus is in the shape of little blueish dice which are composed of chalk or sulphate of lime, impregnated with a blue dye which may be extracted by water. This blue color is reddened by acids; alkalies restore to blue the reddened litmus. Solution of litmus is made by soaking an ounce of litmus in several ounces of water, and filtering. Litmus paper is made by dividing in halves a solution of litmus, and stirring one-half with a glass rod touched repeatedly in a drop of oil of vitriol until the color *just begins* to turn red, and then adding the other half; the object of this is to render the litmus more sensitive to acids, by neutralizing one-half of its natural alkali. Strips of *unsized* paper are dipped in this solution until distinctly blue, drying them in the intervals; this gives "blue litmus paper." To the rest of the solution, enough oil of vitriol is added on a glass rod, with which it is stirred until it is just fairly red; with this the red litmus paper is prepared.

In testing vinegar by this process, *two ounces* of it are weighed in a *beaker* glass, which is a chemical vessel, in shape not unlike a tumbler, and with a thin bottom, so that it may bear the application of heat. The beaker, with the vinegar is placed upon a support, upon two pieces of wire gauze, (to diffuse the heat,) and heated by a spirit lamp underneath. The object of warming is to drive off the carbonic acid, liberated from the carbonate of soda, since this gas is soluble in cold water, of acid nature, and if present would prevent our ascertaining when the exact point of neutralization was reached, by its reddening action upon the litmus paper. A little solution of litmus is now added to the vinegar in the beaker, which at once turns red. It is important not to add too much litmus, as, approaching the point of neutralization it would assume a violet hue, which would be difficult to distinguish from red or blue.

The bottle of carbonate of soda is now counterpoised upon the balance, and small portions of the salt taken from it by a little spoon and added to the warm vinegar, when the effervescence of the carbonic acid is at once perceived. Care must be taken that no liquid spirt upon the little spoon, which would cause the powdered carbonate of soda to adhere to it. Should this accident take place, the spoon must be washed with a little distilled or rain water, (which is added to the vinegar,) and then dried; to prevent some of the vinegar being projected by the effervescence, the beaker should be covered with a glass plate, from which the drops are washed into the vinegar when approaching the point of neutralization. As we near this point, the dyed vinegar begins to experience a change of color; the soda must now be added cautiously, in very small portions, stopping after each addition until the effervescence ceases, and before adding fresh soda, dipping a strip of red and one of blue litmus paper in the solution. As soon as the *blue* paper is not reddened, and when the *red* paper just begins to be slightly blued, the point of neutralization is reached. We now ascertain how many grains must be added to the scale-pan holding the bottle of soda; the requisite number gives the quantity of soda used for neutralization. This number is divided by 27 to get the percentage of *anhydrous* acetic acid in the vinegar, or by 23 to obtain its percentage of hydrated acid.

2d. *By dry Carbonate of Soda.*—This salt may be employed exactly as described for the crystalized carbonate. I will give directly the quantity of it which corresponds to 1 per cent. of acetic acid when operating upon two ounces of vinegar. Dry carbonate of soda is made by heating upon a plate in a *very* hot stove the pure bi-carbonate of soda. By the heat half the carbonic acid and all the water escape, leaving dry neutral carbonate of soda.

3d. *By dry Carbonate of Potassa.*—This salt may also be employed in the same manner. It must be perfectly pure, and is prepared preferably from cream of tartar. Before using, it must be strongly heated, and suffered to cool in a glass stoppered bottle; because it has the property of attracting moisture from the air. This property renders it an unpleasant test for vinegar.

The following table embraces the different alkalies employed in the preceding methods:

TABLE OF VINEGAR TESTS.

Employing 2 ounces, (960 grains) of vinegar.

	For 1 per cent. of acid in the vinegar are required for neutralization.	
	As Anhydrous acetic acid.	As Hydrated acetic acid.
For Crystaline Carbonate Soda,	27 grains,	23 grains.
" Dry " "	10 "	8½ "
" Dry Carbonate Potassa,	13 "	11 "

For example, suppose that thirty-five grains of dry carbonate potassa were taken to neutralize the acid in two ounces of vinegar; to get its per centage of *hydrated* acetic acid, we would divide $\frac{35}{11} = 3\cdot 18$ per cent.

II. METHOD BY WEIGHING ALKALINE CARBONATED SOLUTIONS.

This method is carried out as the preceding ones; but instead of weighing the *dry* salts we weigh their definite solutions. *One* part by weight of crystalized carbonate of soda or of dry carbonate of potassa, is dissolved in *three* parts by weight of distilled or rain water, making solutions containing one-quarter part by weight of salt. As the solution does not change, a sufficient quantity may be made once for all, and kept in well stoppered bottles. The operation is carried on as by salts, but with much greater facility. A small bottle of the solution is counter-balanced, and the quantity needed for any experiment ascertained by loss of weight.

The divisors for getting the per cents. are four times as large as those given for the methods by dry salts. The following tables spare even this small calculation. The tables may serve for the dry process by multiplying the results of the dry process by four.

Examples from the tables. Suppose that 540 grains of the solution of crystalized carbonate soda were needed to neutralize the acid in 2 ounces of vinegar. The vinegar (by Table I.) contains 5% of anhydrous acetic acid, and (by Table II.) not quite 6% of hydrated acetic acid. If 156 grains of solution carbonate potassa were employed, then the vinegar contains (Table I.) 3% anhydrous acetic acid, and (Table II.) 3·5% hydrated acetic acid. Tables III. and IV. are for those who prefer to use the French weights. To employ them, 50 *grammes* of vinegar must be operated upon, and the weight of the solutions necessary for neutralization determined in *grammes*.

VINEGAR MANUFACTURE.

TABLE I.
In 2 ounces (960 grains) vinegar.

Grains employed of solutions containing ¼ salt.		Per centage Anhydrous Acetic Acid.
Of Cryst. Carb. Soda.	Of Dry Carb. Potas.	
108	52	1
135	65	1·25
162	78	1·50
189	91	1·75
216	104	2
243	117	2·25
270	130	2·50
297	143	2·75
314	156	3
351	169	3·25
378	182	3·50
405	195	3·75
432	208	4
459	221	4·25
486	234	4·50
513	247	4·75
540	260	5
567	273	5·25
594	286	5·50
621	299	5·75
648	312	6
675	325	6·25
702	338	6·50
729	351	6·75
756	364	7
783	377	7·25
810	390	7·50
837	403	7·75
864	416	8
891	429	8·25
918	442	8·50
945	455	8·75
972	468	9
999	481	9·25
1026	494	9·50
1053	507	9·75
1080	510	10

TABLE II.
In 2 ounces (960 grains) vinegar.

Grains employed of solutions containing ¼ salt		Per centage of Hydrated Acetic Acid.
Of Cryst. Carb. Soda.	Of Dry Carb. Potas.	
91	44	1
114	55	1·25
137	66	1·50
159	77	1·75
182	88	2
205	99	2·25
228	110	2·50
251	121	2·75
273	132	3
296	143	3·25
319	154	3·50
342	165	3·75
365	176	4
387	187	4·25
410	198	4·50
432	209	4·75
455	220	5
478	231	5·25
500	242	5·50
513	253	5·75
546	264	6
569	275	6·25
592	286	6·50
614	297	6·75
637	308	7
660	319	7·25
683	330	7·50
706	341	7·75
728	352	8
751	363	8·25
774	374	8·50
797	385	8·75
820	396	9
842	407	9·25
865	418	9·50
888	429	9·75
911	440	10

ACETOMETRY.

TABLE III.			TABLE IV.		
In 50 grammes of vinegar.			In 50 grammes of vinegar.		
Grammes of the solutions employed containing ¼ salt.		Per centage Anhydrous Acetic Acid.	Grammes of the solutions employed containing ¼ salt.		Per centage of Hydrated Acetic Acid.
Of Cryst. Carb. Soda.	Of Dry Carb. Potas.		Of Cryst. Carb. Soda.	Of Dry Carb. Potas.	
5·6	2·7	1	4·7	2·3	1
7·0	3·4	1·25	5·9	2·8	1·25
8·4	4·0	1·50	7·1	3·4	1·50
9·8	4·7	1·75	8·3	3·9	1·75
11·2	5·4	2	9·5	4·5	2
12·6	6·1	2·25	10·7	5·0	2·25
14·0	6·7	2·50	11·9	5·6	2·50
15·4	7·4	2·75	13·1	6·2	2·75
16·8	8·1	3	14·3	6·8	3
18·2	8·8	3·25	15·4	7·4	3·25
19·6	9·4	3·50	16·6	8·0	3·50
21·0	10·1	3·75	17·8	8·5	3·75
22·4	10·8	4	19·0	9·1	4
23·8	11·5	4·25	20·2	9·7	4·25
25·2	12·2	4·50	21·4	10·2	4·50
26·6	12·8	4·75	22·6	10·8	4·75
28·0	13·5	5	23·8	11·4	5
29·4	14·2	5·25	25·0	12·0	5·25
30·8	14·9	5·50	26·2	12·5	5·50
32·2	15·5	5·75	27·3	13·1	5·75
33·6	16·2	6	28·5	13·7	6
35·0	16·9	6·25	29·7	14·3	6·25
36·4	17·6	6·50	30·9	14·8	6·50
37·8	18·2	6·75	32·1	15·4	6·75
39·2	18·9	7	33·3	16·0	7
40·6	19·6	7·25	34·5	16·6	7·25
42·0	20·3	7·50	35·7	17·2	7·50
43·4	21·0	7·75	36·9	17·7	7·75
44·8	21·6	8	38·1	18·3	8
46·2	22·3	8·25	39·2	18·9	8·25
47·6	23·0	8·50	40·4	19·5	8·50
49·0	23·7	8·75	41·6	20·0	8·75
50·4	24·3	9	42·8	20·6	9
51·8	25·0	9·25	44·0	21·2	9·25
53·2	25·7	9·50	45·2	21·8	9·50
54·6	26·4	9·75	46·4	22·4	9·75
56·0	27·0	10	47·6	23·0	10

III. METHOD BY MEASURING ALKALINE CARBONATED SOLUTIONS.

By the preceding methods everything was *weighed;* by that about to be described only one weighing is necessary. Employing this method, an analysis of vinegar may be made in a few minutes. I have used in the description the French weights and measures, which are generally employed by scientific men of all countries; but any other weights may be used by maintaining the same proportions. I may remark here, that by the French system the *metre* is the unit of length, and is the ten millionth part of a quarter of the earth's meridian. It is divided into tenths, (decimetres,) hundredths, (centimetres cubes,) and thousandths, (millimetres). Compared with our measure, the metre equals 39·371 inches. The weight of the volume of one cubic centimetre of distilled water at 39° Fah. is the gramme; it corresponds to 15.434 of our grains. If we desire to measure the volume of 1 cubic centimetre in a tube, we have only to weigh therein one gramme of distilled water, bring it to 39° Fah., and mark its level on the tube. The litre is 1000 cubic centimetres.

Figure 2 represents a litre measure; a flask which filled to the mark on the neck, will measure one litre very exactly. It may be made by counterpoising a suitable flask, and weighing into

ACETOMETRY. 183

Fig. 2.

it of pure distilled water of temperature 39° Fah. 1000 grammes or 15434 grains, which equal 2 pounds 8 ounces 1 drachm 14 grains, apothecaries' weight. The weights of course must be accurate. This litre measure is employed for making a test solution of carbonate of soda of definite strength, as follows: On the supposition that two weighed ounces of vinegar be always taken for analysis; then weigh 2690 grains of *crystalized*, or 1000 grains of *dry* carbonate of soda. This weight may be made once for all of brass or even of block tin, and given to an apothecary to weigh the soda with, if the vinegar maker, unfortunately, does not possess a balance.

Introduce the soda into the litre flask, which must then be ⅔ filled with rain or distilled water, and placed in a warm situation, occasionally agitating it, but not getting any of the liquid upon the neck, until the salt be completely dissolved. When the solution is *perfectly cold*, add pure water to the mark, and shake the flask well. Observe that 10 cubic centimetres of this solution contain 27 grains of crystalized, or 10 grains of dry carbonate of soda, which, in saturating two ounces of vinegar, is equal to *one per cent.* of anhydrous acetic acid. If it be desirable to indicate the acetic acid in the hydrated state, then, in preparing the test solution, use 2280 grains of crystalized, or 850 of dry carbonate of soda. Operating upon 2 ounces of vinegar, 10 cubic centimetres will indicate 1 per cent. of hydrated acetic acid.

The analysis is performed as described for Method II., except that the test solution is poured from a *measure* graduated in *cubic centimetres.* The number of cubic centimetres of the test solution requisite for neutralizing 2 ounces of vinegar, if divided by 10, gives the acid per centage without further calculation. *Example.* Suppose 48 cubic centimetres were required, then $\frac{48}{10} = 4\cdot 8$ per cent. of acetic acid in the liquid tested.

If we analyze 1 ounce of vinegar, (480 grains,) the number of cubic centimetres, divided by 5, gives the per centage of acid.

The vinegar employed for the analysis need not be weighed; it can be *measured,* which will enable this test to be carried on without any weighing at all, having once prepared the test solution.

The measure is made by weighing 2 ounces of vinegar (of the strength usually tested) in a small

ACETOMETRY. 185

flask, shaped like the litre flask, making a mark at the level on the neck. A more convenient means is the pipette, fig. 3.

Fig. 3. Having a glass pipette large enough to hold 2 ounces of vinegar rising to a level in the upper stem, fill it with vinegar, let the same drain out, and close the lower aperture with a little pellet of wax; then counter balance it exactly, and weigh into it two ounces of vinegar, marking the level to which the liquid rises in the stem. With this instrument, it is easy to measure accurately two ounces of vinegar, by introducing the lower end into the vinegar, and applying suction at the upper end, until the liquid rises above the mark. The pipette is then removed with the finger closing the upper aperture, and air admitted until enough vinegar drops to bring the level of the liquid to the mark.

The graduated measure for the test solution is called a "*burette.*" There are several varieties of this instrument.

Here are two kinds. Figure 4 explains itself.

If the glass tube be reasonably cylindrical the graduation may be made by weighing successively into it portions of 10 grammes (154·34 troy grains) of distilled water to obtain the points 0—10—20 —30, &c., cubic centimetres, and by dividing the spaces thus obtained into ten degrees.

Fig. 5 represents Mohr's burette, which has found universal favor with chemists, and can be

VINEGAR MANUFACTURE.

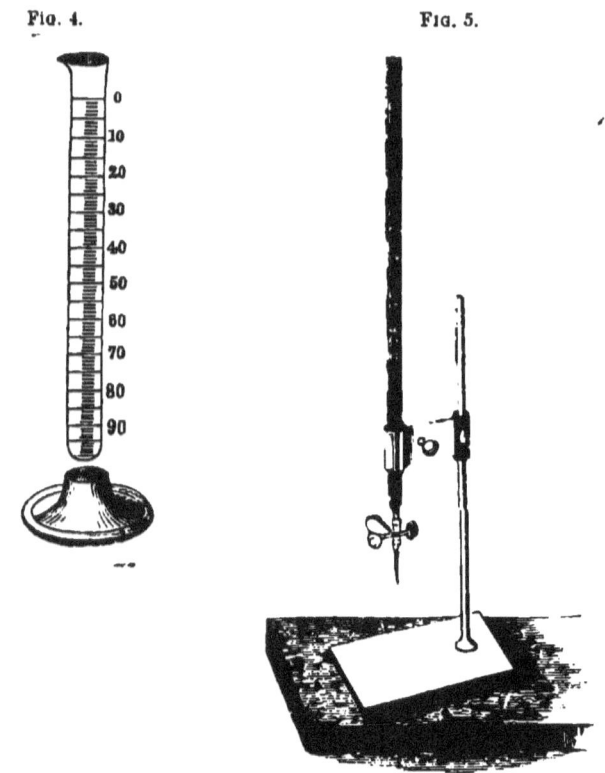

Fig. 4.   Fig. 5.

recommended as the best. It consists of a tube graduated from above downward; open above, and drawn out below cylindrical, of the thickness of a straw for half an inch, and then constricted to an opening sufficiently fine to permit a liquid to pass in a very thin stream. Mohr's invention consists in a *spring clip*, or "*compressor*," which enables an admirable cock for chemical purposes to be made from a piece of vulcanized rubber tube.*

---
* Quéttier, 193 Greenwich Street, New York, imports from France the only article of vulcanized tubing really suited to chemical use.

Fig. 6.

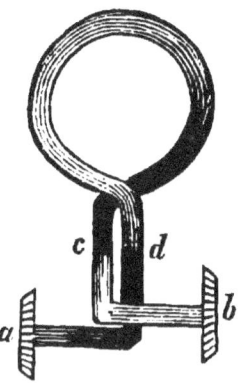

Figure 6 represents the compressor. It is made of steel, well tempered. By pressing the buttons, $a\ b$, the jaws, $c\ d$, are opened for the insertion of the rubber tube, which is squeezed flat on releasing the jaws. When it is required to permit a passage of gas or liquid through the tube, the buttons are compressed to the required degree. This compressor may be applied to either end of the burette. If applied, as in the figure, the vulcanized tube is tied to the half-inch cylindrical portion; then, sufficient space being allowed for the compressor, a small bit of glass tube drawn to so fine an opening that liquid will remain in it by capillary attraction when the compressor is shut, is tied to the other end of the rubber tube.

To fill the burette, a vessel containing the test liquid is placed below, and releasing the compressor, suction is applied to the top of the burette to fill it above the 0° mark. Enough liquid is then permitted, by manipulating the compressor, to escape until the level stands accurately at 0°, care being taken that no air lurks in the lower part of the instrument.

When the compressor is applied above, it pinches a long piece of vulcanized rubber tube tied to the upper end of the burette, which then of course contains no addition at the lower end. It is filled and manipulated as before, and when the compressor is closed, exclusion of the atmospheric pressure from above keeps the liquid in the tube.

In the foregoing description, I have employed grain weights for the analysis, and French weights and measures for graduating the instruments. The reason of this apparently incongruous union is, because excellent litre flasks and cubic centimetre graduates may be purchased ready made at the philosophical instrument makers, having which nothing is required but apothecaries weights in carrying on the analysis as described. As I have given the equivalents of grammes in grains in describing the graduates, it is easy to make them with apothecaries weights.

Finally, if gramme weights are employed altogether, the test liquid is prepared in the litre flask by taking 104 *grammes* of *dry*, or 280 grammes of crystalized carbonate of soda, and by operating upon 50 *grammes* of vinegar. The measuring pipette for the vinegar will then be made by weighing in it 50½ grammes of water, and marking its level. The volume of the additional half gramme of water is on account of the superior density of the vinegar; 50½ cubic centimetres of vinegar of the strength usually examined, will weigh about 50 grammes.

Five cubic centimetres (Burette degrees) of this test liquid will denote 1 per cent. of *anhydrous* acetic acid.

To prepare the test solution to denote a per centage of hydrated acetic acid; take 82·25 grammes of dry, or 238·3 of crystalized carbonate of soda.

IV. THE AMMONIA PROCESS BY OTTO'S ACETOMETER.

The principal difficulty which the inexperienced have to contend with in the foregoing methods, arises from the acid re-action of the carbonic acid evolved from the alkaline carbonates employed, rendering the point of saturation, which is judged of by the color of the litmus, uncertain. Otto has obviated this difficulty by the employment of aqua ammonia for the neutralizing or test liquid, and by the invention of a simple graduated tube, (which he calls an acetometer,) in which the whole analysis is performed by the most inexperienced person. When the test solution is once prepared, neither weights nor calculation are necessary to perform, in a few minutes, an analysis of vinegar.

Figure 7, illustrates the acetometer. It is a truly cylindrical glass tube of ½ inch bore, 12 inches long, and closed at the end, *a*. It is graduated in the following manner: *a* denotes the

Fig. 7. point of level of exactly *one* gramme of water. Between *a* and *b* it contains *ten* grammes of water, between *a* and *x five* grammes of water. The volumes *b c*, *c d*, *d e*, &c., are obtained by the addition successively of 2·08 grammes of water, if the acetometer shall indicate per centage of hydrated acetic acid, and of 2·447 if it shall indicate anhydrous acetic acid. These volumes are taken, because 2·08 grammes of water occupy the same space as 2·07 grammes of aqua ammonia, containing 1·369 per centage of pure ammonia. This quantity saturates $\frac{1}{10}$ of a gramme of *hydrated* acetic acid; 2·447 grammes of water occupy the volume of 2·435 aqua ammonia of the above strength which saturates $\frac{1}{10}$ gramme of *anhydrous* acetic acid.

The tube should be marked, "For hydrated acetic acid," or, "For anhydrous acetic acid," according to which kind of per centage it is graduated for. The volumes *b c*, *c d*, *d e*, &c., may be each sub-divided into 4 or 8 equal parts.

To employ the acetometer are required; an ammonia test solution, the preparation of which will be given in detail; also, a solution of litmus made by suffering four ounces of water to stand for some time upon a drachm of litmus, then filtering off the sediment.

The analysis is performed in the following simple manner, neglecting the point *x* for the present:

1. Fill to *a* with litmus solution.
2. Fill to *b* with the vinegar under examination.
3. Add the ammonia test until the *red* color of the liquid just begins to turn blue.
4. The level which the liquid now occupies will designate the per centage. For example, suppose the tube is graduated for anhydrous acetic acid, and the level of liquid at the close of the operation, stand at *g;* the vinegar then contains 5 per cent. anhydrous acetic acid. In order to obtain accurate results, the following precautions are to be taken:

1. After pouring the litmus let the tube rest awhile to see whether the level is exactly at *a*; if not, add or take away enough to bring it to this mark. A glass tube of the thickness of a straw, drawn to a fine opening, is useful in regulating this level.

2. Use greater precaution in adding the vinegar, pouring first nearly enough to bring the level to *b*, and the rest drop by drop. An excess of vinegar cannot be removed, for it is now diluted by the litmus water.

3. The acetometer should have a thin edge, and be readily closed by the thumb.

4. The ammonia solution may be added by a dropping tube, (which, if large and stationary, may have compress cock,) or it may be dropped from a bottle with a *thin* lip.

5. After each addition of ammonia close the acetometer accurately with the thumb, holding it

in the left hand, and mix the solution by inverting it a couple of times, then scrape the liquid from the thumb on the thin edge of the acetomeeter, so that none be lost. When neutralization takes place, let the instrument stand a short time before reading its indication, so that the liquid on the sides flows down into that in the tube.

6. If the strength of vinegar be known within $\frac{1}{2}$ per cent., that quantity of the ammonia may be at once added and mixed, and the remaining portion necessary for neutralization then added very carefully.

7. Stop when the color has fairly changed to blue.

8. A second experiment with the same vinegar will enable great accuracy to be attained, since we know when we are arriving at the neutralization point, and can proceed with due caution.

The point x, is so situated, that $a\,x$ contains 5 grammes of water, and is for the purpose of analyzing a vineger stronger than 12 per cent., for which the instrument is graduated. If the vinegar be stronger than 12 per cent., proceed as before, but add the vinegar to x only, and *pure water* to b, and *double* the degrees read after neutralization. If the vinegar be very weak, then add an equal volume of water to as much of the ammonia test solution as may be required, and proceed as in the first example, taking *half* the indication to get the per centage. If the acetometer be constructed for per centage of hydrated acetic acid, its corresponding strength may be ascertained in anhydrous acid, or *vice versa* by employing the tables on pages 169, 170.

PREPARATION OF THE AMMONIA TEST SOLUTION.

Much of the accuracy of Otto's method depends upon the ammonia solution employed for neutralization. It is, therefore, necessary to devote some space to a detailed description of the manner of preparing it. It may be furnished by a reliable apothecary,* using the directions in this chapter; but even then the vinegar maker should possess the knowledge requisite for testing its quality. Take a given quantity, say a pound, of the purest aqua ammonia of commerce. To be able to dilute this to the strength for testing, namely, until it contains 1·369 per centage of ammonia, we must first ascertain accurately how much of this volatile alkali it contains, which is effected by taking its specific gravity.

Any of the following methods may be employed:

1. By the specific gravity, bottle as described on page 133. This is the best method.

2. By a delicate hydrometer, indicating specific gravities between 0·951 and 0·978 made for the purpose.

* Messrs. Bullock & Crenshaw, of Philadelphia, can be recommended to those desiring to purchase pure chemical re-agents. I name this firm from personal knowledge, and with no desire to detract from the merits of others.

17

3. If neither of the above are at hand, an accurate Tralles alcoholometer, (*i. e.*, one of which the degrees indicate alcoholic percentage by volume,) may be used, taking care that the temperature of the ammonia is exactly 62° Fah., and that the degrees are read with great precision. The specific gravity may be ascertained by the following table.

DEGREES OF TRALLES' ALCOHOMETER CONVERTED TO SPECIFIC GRAVITIES.

TRALLES DEGREES.	SPECIFIC GRAVITIES.
25	0·970
26	0·969
27	0·968
28	0·967
29	0·966
30	0·965
31	0·964
32	0·963
33	0·962
34	0·961
35	0·960
36	0·958
37	0·957
38	0·955
39	0·953
40	0·952
41	0·950

Having determined the specific gravity, the test solution may readily be made by means of the data in the following table.

ACETOMETRY.

Aqua Ammonia.		To make 1000 parts by weight of a test solution, containing 1·369 per cent. of ammonia. Take	
Which contains the following percentages of ammonia.	Has the following specific gravities	Aqua Ammonia.	Water.
12·000	0·9517	114·8	885·2
11·875	0·9521	115·3	884·7
11·750	0·9526	116·5	883·5
11·625	0·9531	117·8	882·2
11·500	0·9536	119·0	881·0
11·375	0·9540	120·0	880·0
11·250	0·9545	121·7	878·3
11·125	0·9550	123·0	877·0
11·000	0·9555	124·5	875·5
10·954	0·9556	125·0	875·0
10·875	0·9559	126·0	874·0
10·750	0·9564	127·3	872·7
10·625	0·9569	129·0	871·0
10·500	0·9574	130·4	869·6
10·375	0·9578	132·0	868·0
10·250	0·9583	133·5	866·5
10·125	0·9589	135·0	865·0
10·000	0·9593	137·0	863·0
9·875	0·9597	138·6	861·4
9·750	0·9602	140·4	859·6
9·625	0·9607	142·2	857·8
9·500	0·9612	144·0	856·0
9·375	0·9616	146·0	854·0
9·250	0·9621	148·0	852·0
9·125	0·9626	150·0	850·0
9·000	0·9631	152·0	848·0
8·875	0·9636	154·0	846·0
8·750	0·9641	156·4	843·6
8·625	0·9645	158·7	841·3
8·500	0·9650	161·0	839·0
8·375	0·9654	163·5	836·5
8·250	0·9659	166·0	834·0
8·125	0·9664	168·5	831·5
8·000	0·9669	171·0	829·0
7·875	0·9673	173·8	826·2
7·750	0·9678	176·6	823·4
7·625	0·9683	179·5	820·5
7·500	0·9688	182·5	817·5
7·375	0·9692	185·6	814·4
7·250	0·9697	188·8	811·2
7·125	0·9702	192·0	808·0

VINEGAR MANUFACTURE.

Aqua Ammonia.		To make 1000 parts by weight of a test solution containing 1·369 per cent. of ammonia. Take	
Which contains the following percentages of ammonia.	Has the following specific gravities.	Aqua Ammonia.	Water.
7·000	0·9707	195·6	804·4
6·875	0·9711	199·0	801·0
6·750	0·9716	202·8	797·2
6·625	0·9721	206·6	793·4
6·500	0·9726	210·6	789·5
6·375	0·9730	214·7	785·3
6·250	0·9735	219·0	781·0
6·125	0·9740	223·5	776·5
6·000	0·9745	228·0	772·0
5·875	0·9749	233·0	767·0
5·750	0·9754	238·0	762·0
5·625	0·9759	243·4	756·6
5·500	0·9764	249·0	751·0
5·375	0·9768	254·7	745·3
5·250	0·9773	260·8	739·2
5·125	0·9878	267·0	733·0
5·000	0·9783	273·8	726·2

The use of the table is very simple. Suppose the specific gravity of the aqua ammonia be ascertained to be 0·965; find this number in the second column, and on consulting the numbers opposite in the 3d and 4th columns, it will be seen that to make one thousand parts of the test solution, we must take 161 parts, by weight, of this ammonia, and mix with 839 parts by weight of pure water. The first column merely indicates the percentage of ammonia corresponding to the respective specific gravities.

We are now able to dilute our pound of ammonia to the required degree; but we must be sure that it is an accurate pound, and must know what kind of a pound it is; for in our country

ACETOMETRY. 197

we weigh, with three different pounds: Troy, Apothecaries' and Avoirdupois.

AVOIRDUPOIS WEIGHT.

	Ounces.		Drachms.		Troy grains.
℔ 1	=	16	=	256	=	7000
	oz. 1	=	16	=	437·5
			dr. 1	=	27.34

APOTHECARIES' WEIGHT.

	Ounces.		Drachms.		Scruples.		Troy grains.
℔ 1	=	12	=	96	=	288	=	5760
	℥ 1	=	8	=	24	=	480
			ʒ 1	=	3	=	60
					℈ 1	=	20

The Imperial Standard Troy weight recognized by the British Government, corresponds with Apothecaries weight in *pounds, ounces* and *grains*, and differs only in the *sub-division* of the ounce. In Troy weight, 24 grains = 1 pennyweight, dwt. and 20 pennyweights, = 1 ounce.

If the ammonia weighed one pound by Avoirdupois weight, and had a specific gravity of 0·9650, we must then dilute 7000 grains, *i. e.*, a pound of ammonia, with water, in the proportion of 161 ammonia to 839 water, or by the rule of three: 161 : 839 :: 7000 : $x = \frac{839 \times 7000}{161} = 36478$ grains, which is equal to 5 lbs. 3 oz. and 167 grains of water to be added to a pound of the ammonia. The test liquid may be preserved in *small* bottles with accurately fitting glass stoppers. After filling the bottles, wipe the inside of the neck; and having inserted the stoppers with moderate friction, pour melted wax around the cavity between the neck and the stopper, and tie moist bladder over the stopper. A bottle may be opened as occasion requires. A few pounds of this test solution will keep well and last for a considerable length of time.

Another method of preparing the test solution:
By this method the determination of the specific gravity of the ammonia is avoided. The strength of the ammonia solution is ascertained by the acetometer, by means of a solution of known strength, either of acetic or tartaric acid.

A. *By acetic acid.*—A vinegar of any known strength below 12 per cent. may be employed, but it is better to prepare a solution of pure acetic acid for the purpose, for if well bottled it will keep, and may be made use of at any future time for making ammonia test solution. The acidum aceticum of the United States Dispensatory has a specific gravity of 1·06, and contains 40 per cent. of anhydrous acid; consequently, if we add, by weight, three times as much water as we take acid, we will have a 10 per cent. solution. It will not do, however, to take the strength of the acidum aceticum on trust, but test the resulting 10 per cent. solution, by the specific gravity bottle, or by the hydrometer, to see whether it has its corresponding specific gravity, viz: 1·0177 at 59° Fah. Having thus a vinegar of known per

ACETOMETRY.

centage, we next take a pound of ammonia and mix it with 4 or 5 pounds of rain or distilled water. With this, perform the analysis of the known vinegar in the acetometer, as described on page 190, on the supposition that the known vinegar contains 10 *per cent* of anhydrous acid; if the ammonia solution is correct, which can only be accidentally the case, the acetometer will give an indication of 10 per cent; if it give more, add some *strong ammonia* to the ammonia, and repeat the analysis (adding strong ammonia each time) until it gives less than the known per centage of the acid, when the *difference* between *the known* and *the indicated* per centage will inform us how much water to add to the ammonia to bring it to the standard proper for testing. For example, we *know* that the vinegar contains, in this case, 10 per cent. of acid; if the acetometer give an indication of $9\frac{1}{2}$ per cent., we have to add 1 per cent. of water to make these $9\frac{1}{2}$ degrees, 10. If, then, we take $9\frac{1}{2}$ pounds of the solution of ammonia, and add 1 pound of water, it will yield a solution of the strength required to use with Otto's acetometer. Repeating the analysis with this, we will find that the acetometer indication will be 10°, corresponding to the known per centage of our vinegar. Again, suppose the known vinegar contains $6\frac{1}{2}$ per cent. acid, and we had an acetometer indication of $5\frac{3}{4}$ per cent.; then to every $5\frac{3}{4}$ pounds of the ammonia we must add $\frac{3}{4}$ pound of water, for $5\frac{3}{4} + \frac{3}{4} = 6\frac{1}{2}$.

B. *By Tartaric Acid.*—This acid, *if pure*, may be very conveniently employed instead of acetic acid of known strength, for preparing the ammonia test solution. Take pure dry crystals of tartaric acid, powder them, and preserve after pressing between sheets of porous paper in a stoppered bottle.

1.47* grammes of tartaric acid have the same neutralizing power as the quantity of ten per cent. (anhydrous) acetic acid used in the Acetometer. The litmus solution having been introduced into the acetometer, the above quantity of tartaric acid is added, (which an apothecary may weigh upon his balance with the above weight,) and then water to the mark b; see fig. 7. The acid is dissolved by gently agitating the acetometer.

Proceed with the test as in A, and interpret the result in the same manner. The difference between 10 and the indicated per centage, will give the amount of water to be added to the ammonia. For example, suppose the indication were 8 per cent., then to every 8 pounds, ounces, &c., of the diluted ammonia, we must add 2 pounds, ounces, &c., of water because $8+2=10$.

Fault having been found with this process, from the fact that neutral acetates of the alkalies render blue reddened litmus paper, and consequently the per centage of any vinegar must be given too low with the instrument; Otto performed a series of careful experiments, and ascertained that;

* A weight of this kind should be sold with the Acetometer.

1st. The error in question does not exceed on an average $\frac{1}{10}$ of one per cent.

2d. That it is counterbalanced by the fact that in practice always a *little more* ammonia is added than for exact neutralization.

3d. If the vinegar be measured by *volume*, as when the acetometer is graduated by the volume *a b* of 10 grammes of distilled water, its weight is taken about 1 per cent. higher than it really is by reason of the difference between the specific gravities of water and vinegar. The result of this opposition of errors is that even in unpracticed hands, Otto's method is extremely correct, if the acetometer be accurate, and the ammonia solution properly made.

For testing vinegar by this method, the philosophical instrument maker should prepare a box containing the necessary apparatus, as;

1st. The acetometer graduated for per centages of "anhydrous acetic acid," with these words marked on the instrument.

2d. A weight of 1.47 grammes.

3d. A bottle of litmus solution (1 drachm litmus+4 ounces water) and one of solid litmus for preparing the solution.

4th. A bottle of pure pulverized tartaric acid.

5th. A bottle containing pure solution of acetic acid containing 10 per cent. anhydrous acid.

6th. A bottle containing pure aqua ammonia of the correct strength for testing.

All of the bottles should have *thin* lips, that they may drop well.

It would be well to add a correct but not expensive balance and weights.

Such a case of apparatus would not be costly and would place the vinegar maker, or other interested person, in the position of performing an accurate analysis of vinegar in a few minutes.

V. *Balling's Vinegar Test.*—This test determines the acid per centage of vinegar by observing the *increase* of specific gravity of the liquid undergoing acetification. The process is an analogous one to the information of alcoholic strength gained by watching the attenuation of worts by the saccharometer. Balling has invented an instrument for this purpose, which, improved by Otto, is constructed in the following manner. It is a very delicate hydrometer, *i. e.*, one with big body and slender stem, which is graduated thus: Three points are marked on the stem (at a temperature of 63° Fah.) viz., a central one to where it sinks in water; a superior one to where it sinks in 10 per cent. alcohol, which has a specific gravity of 0.9841; and an inferior one to where it sinks in 10 per cent. vinegar, specific gravity of 1.034. The length on the stem between the central and superior points is divided into 5 equal parts. The length between the central and inferior point, into 10 equal parts. We have thus 15 degrees which may be further subdivided, and which are numbered from the top downward, placing 0° at the superior point. In this instrument every degree indicates an increase of 0.0033 in specific gravity, or one per cent. anhydrous acetic acid.

If this hydrometer be placed in a liquid of which the specific gravity is 0.9841, and *increasing*, it will first sink to 0°, and as its specific gravity increases the stem will rise from the liquid until 5 degrees have passed, when the liquid will have the same density as water. It will continue to rise until 10 more degrees have passed, when the liquid will have the specific gravity of 10 per cent. acetic acid.

Its use is very simple. Before acetification bring the alcoholic mixture to a temperature of 63° Fah., and mark the indication of the (Balling) hydrometer. Suppose it to be $2\frac{1}{2}$°. After acetification, bring the vinegar to 63° Fah., and observe again the indication of the instrument. Suppose it to be 6. Then, since every degree gained denotes an increase of 1 per cent. anhydrous acetic acid, the per centage of the vinegar in question is found by simply taking the difference of the two indications. In the example the per centage is 6 less $2\frac{1}{2}$ equals $3\frac{1}{2}$.

THE ACID STRENGTH OF THE VINEGAR COMPARED WITH THE ALCOHOLIC STRENGTH OF THE MIXTURE.

The following tables are useful to the vinegar maker.

VINEGAR MANUFACTURE.

TABLE I.—FOR WEIGHTS PER CENT. OF ALCOHOL.

Vinegar mixture of		Yields		Equals Vinegar.	Per centage of vinegar in anhydrous acid.
Alcohol.	Water.	Anhydrous acetic acid.	Water.		
1	99	1·108	99·587	100 695	1·100
2	98	2·216	99·174	101·390	2·185
3	97	3·324	98·761	102·085	3·251
4	96	4·432	98·348	102·780	4·312
5	95	5·540	97·935	103·475	5·354
6	94	6·648	97·522	104·170	6·382
7	93	7·756	97·109	104·865	7·397
8	92	8·864	96·696	105·560	8·399
9	91	9·972	96·283	106·255	9·385
10	90	11·080	95·876	106·950	10·360

TABLE II.—FOR VOLUMES PER CENT. OF ALCOHOL.

A mixture containing the following per cents of alcohol by volume.	Is composed by weight of		And yields		Total vinegar.	Per centage of anhydrous acetic acid.
	Alcohol.	Water.	Acetic acid.	Water.		
1	0·795	99·205	0·881	99·671	100·552	0·876
2	1·592	98·408	1·764	99·342	101·106	1·744
3	2·392	97·608	2·650	99·012	101·662	2·607
4	3·195	96·805	3·540	98·680	102·220	3·463
5	3·995	96·005	4·426	98·350	102·766	4·306
6	4·804	95·196	5·323	98·066	103·389	5·147
7	5·613	94·387	6·219	97·681	103·900	5·985
8	6·422	93·578	7·115	97·348	104·463	6·811
9	7·234	92·760	8·015	97·012	105·027	7·631
10	8·047	91·953	8·916	96·676	105·592	8·439
11	8·865	91·135	9·822	96·338	106·160	9·252
12	9·680	90·320	10·725	96·002	106.727	10·049

Example, Table I. If the alcoholic mixture with which the vinegar is to be made contain 6 per cent. by weight of alcohol, 100 lbs. of it will yield 104.170 lbs. of vinegar, composed of 6.648

lbs. of anhydrous acetic acid, and 97·522 lbs. of water. Such vinegar has a per centage of 6·382 anhydrous acid.

By Table II.—If the alcoholic mixture contain 7 per cent. by volume of absolute alcohol, then 100 lbs. of it will contain 94·387 lbs. of water and 5·613 lbs. alcohol, and will yield 103·9 lbs. of vinegar, containing 97·681 lbs. of water, and 6·219 of anhydrous acetic acid, which corresponds to a vinegar of per centage of 5·985 anhydrous acetic acid.

These tables express the *theoretical* quantity of vinegar *possible* from the respective alcoholic mixtures.

In practice this strength is never attained, because, 1st, a certain portion of the alcohol and of the acetic acid are lost by evaporation; and 2d, some of the alcohol remains unchanged at the close of the operation.

The aim of the vinegar manufacturer is to approach these theoretical results as closely as possible. In making vinegar, therefore, of a required per centage strength, it is necessary to take a stronger alcoholic mixture than the tables indicate.

Thus, for a vinegar of from 4·6 to 4·8 per cent. anhydrous acetic acid, an alcohol of 6° Tralles alcoholometer, *i. e.*, 6 per cent. by volume must be employed.

PART II.

PRACTICAL.

CHAPTER I.

GENERAL DETAILS.

HAVING in the first part of this work treated in detail the theoretical points involved in the vinegar process, the practical part of the manufacture may be developed in a much less space.

By far the greater portion of vinegar consumed in our country, is made by the "quick process," to describe which is in fact the object of this book. For the sake of greater completeness, however, it will be necessary to treat briefly the practical details of the old slow process.

Some general principles of detail affecting the acetous transformation; the construction of the factory building, and the kind of water and alcohol employed belong to both processes, and may be advantageously discussed in the present introductory chapter.

Let me, therefore, recall from Part I. a few of the general principles of the acetous transformation.

I.—1st. Vinegar always arises from the transformation of alcohol to acetic acid under the following circumstances: The alcoholic mixture must be *weak;* the temperature between 74–86° Fah.; air must be present and also a ferment.

2d. The higher the temperature and the greater the quantity of air brought in a given time in contact with the alcoholic mixture, the quicker is its transformation to vinegar. If the temperature be too high, alcohol is wasted by evaporation; if too low, the acetification ceases and putrefaction sets in.

3d. One hundred pounds of absolute alcohol are capable of yielding one hundred and eleven pounds of *anhydrous,* (equivalent to one hundred and thirty pounds of *hydrated,*) acetic acid. Consequently, the richer in alcohol the mixture, the stronger in acid will be the vinegar. But note that in *strong* alcoholic washes,* the vinegar transformation is *retarded.*

4th. The more *ferment* present, the quicker is the vinegar process. Vinegar is the best ferment. Certain others, which are nitrogenized bodies, are employed. These are, bread soaked in vinegar, leaven, brewer's yeast, a small portion of dough made of wheat and rye flour, tartar and vinegar. These are used in very small proportions, a few ounces to a barrel of wash. But note well, that the greater the quantity of these last ferments present, the more apt the vinegar is to spoil. They also ruin the shavings in the quick process, by forming putrefying deposits upon them, and by rendering them mouldy.

* The word "*wash*" or "*mixture*," is, technically, the alcoholic solution which is to be made into vinegar.

5th, Saccharine and starchy bodies are capable of being made into vinegar, by first undergoing transformation. The sugars must first, by fermentation, become alcohol, and one hundred pounds of sugar are capable of yielding fifty pounds of absolute alcohol. The starch must first become sugar by the mashing process, and then alcohol by fermentation; one hundred pounds of starch yield one hundred pounds of sugar, which gives fifty pounds absolute alcohol. The alcoholic fermentation requires the presence of a certain ferment, and the conditions of a certain temperature. Some very important considerations arise from No. 5., which may well arrest our attention. Some manufacturers, with a view to increase the acid of the future vinegar, add to their mixture beer, syrup, honey, extract of raisins, juices of sweet fruits, and a variety of other saccharine substances. Some add these to the manufactured vinegar, that it may strengthen by age. These additions are wrong, and proceed from an imperfect knowledge of the principles of the vinegar process. They not only by their sweetness mark the acid taste of the vinegar, but injure its keeping properties. Vinegar is a weak solution of acetic acid containing foreign matters. The weaker it is the more apt it is to spoil. The spoiling is also dependent upon the nature and quantity of the foreign substances which it contains. The lowest grades of vinegar contain 2 per cent.; the better grades from 3 to 6 per cent.;

and the best wine vinegar as high as 10 per cent. of acid. Frequently, in new vinegar, innumerable animalculæ, (the vinegar eels) may be seen swimming. During the process of manufacture, and also when the product is stored, a swollen slippery substance is generated. It is a plant, (Mycoderma aceti,) and is called popularly "mother of vinegar," because it acts as a powerful vinegar ferment, which property is due, doubtless, to the immense amount of vinegar with which it is saturated. These and other foreign substances give, especially to weak vinegar, a tendency to mould and putrefy. An added saccharine solution can never become vinegar without first experiencing the alcoholic fermentation, which requires its own ferment of the nature of yeast. Without this ferment, a portion of the sugar remains unchanged, the remainder gives rise to slimy products and increases the "mother." If yeast be added, or, as in the case of juices of sweet fruits, be generated, certain injurious solid and liquid nitrogenized products arise in the vinegar, injuring its keeping qualities, and forming deposits, (in the quick process,) upon the shavings, which are thereby spoiled. When added to the vinegar of storage, the alcoholic transformation is very slow on account of the low temperature, and the same injurious products are formed in the vinegar. Without doubt the addition of saccharine with yeasty matter accelerates the speed of the vinegar manufacture; but their use is attended by the inconveniences described. Such, therefore, should always be well fermented before being added to the wash.

6th.. The addition of pure alcohol increases the keeping quality of stored vinegar. Vinegar by the quick process generally contains a portion of alcohol; if not, it is well to add it in the proportion of a pint of spirits or whiskey to the barrel of vinegar, which is said technically to live upon the alcohol. In fact, the alcohol in the casks at their low temperature, is converted very slowly into acetic acid, and the vinegar is not so apt to spoil, (other things being equal,) until the alcohol is all gone. It is well, from time to time, to add spirits to vinegar long in store, to preserve it. As elevated temperatures hasten the decomposition of vinegar, it is best stored in cool but not mouldy places.

II. *The "wash" or "mixture."*—The nature of the alcoholic liquid which is to be converted into vinegar, will depend upon the locality of the factory. In some countries it is wine; in others, cider; in our country, in most places, whisky is employed; while in England, owing to the Excise Laws, malt wine is preferred. It is generally more profitable to employ the quick process for whisky or pure alcoholic mixtures, as the shavings are gradually deteriorated by the slow deposits of foreign matter upon them from wine or beer mixtures.

It must be remembered that pure alcohol and

water give a colorless vinegar, and destitute of those agreeable aromatic substances which give value to fine vinegars. Spirits containing their natural fusel oil will yield a vinegar of more pleasant flavor. Indeed, it is sometimes advisable to add a little, but not too much, fusel oil to the spirits with which the wash is made; it is decomposed during the vinegar process, giving rise to pleasant smelling ethers. Vinegars, whether of pure alcohol or whisky, must be colored with a harmless substance, as burnt sugar, to give them the appearance which we naturally expect in vinegar.

III. *The water.*—Pure water is an important requisite for the vinegar manufacturer. The earthy salts of hard water retard the vinegar process, and the carbonates of lime, magnesia, &c., neutralize a certain proportion of the acid. Water that contains organic matter putrifies, and yields a vinegar of inferior keeping qualities. The iron of other waters unites with the tannin of the barrels in which the vinegar is stored, giving it an inky color.

The softer the water the better it is for making vinegar; hence, pure soft and clear spring or river water is the best, next purified rain water, and lastly purified well or spring water.

The vinegar manufacturer may apply to the water he expects to employ, the following simple tests. The substances named are dissolved in pure water and filtered. The water to be tested is placed in wineglasses, and the tests added and stirred in with a slip of glass. On adding the tests a cloudiness, or certain sediments or precipitates occur, from which the nature and hardness of the water may be inferred.

1. *Tincture of Soap.*—Prepared by dissolving a little Castile or similar soap in *weak* alcohol. Occasions a cloudiness, owing to lime, magnesia, iron, &c. The degree of hardness of the water may be inferred from the quantity of the deposit. Hard water, as is well known, does not dissolve soap readily.

2. Solution of carbonate of soda, (common soda,) precipitates the lime from any sulphate of lime, or other soluble lime salt present in water.

3. Solution of oxalic acid effects the same purpose.

4. Solution of chloride of barium precipitates sulphuric acid from soluble sulphates (sulphate of lime, &c.,) in the water.

5. Nitrate of silver (lunar caustic) indicates the presence of chlorides (common salt, &c.,) by a white cheesy precipitate, which becomes black in the light.

6. Solution of gall nuts colors iron waters black.

7. When hard water is boiled, the carbonic acid gas, holding carbonates of lime, magnesia, and iron in solution, is driven off, causing the precipitation of these salts. The amount of this sediment gives a fair indication of the relative

hardness of the water. Sulphate of lime is not precipitated by boiling, unless much water is evaporated.

There are three methods of improving hard water.

1. By boiling; which separates the lime, magnesia, and iron, held in solution by carbonic acid. After settling, the water may be drawn off in a purified condition. Sulphuretted hydrogen is also got rid of thus. This method is costly, and can only be employed under certain favorable circumstances.

2. By exposure to the air, which produces the same effect as 1, but more gradually. The exposure of rivers to the atmosphere is one cause of such water being purer than the spring water of the same localities, which supplies, in a great measure, the rivers. As the carbonic acid gradually escapes into the air, the before mentioned salts are deposited.

3. By filtering. Charcoal in a filter deprives water of certain injurious organic substances, and late discoveries have shown that certain mineral salts are partially removed by the material of water filters.

Filtering should always be performed upon waters that are not perfectly limpid, as their sediment will in the course of time accumulate upon the shavings to a sufficient extent to deteriorate them. Rain water is excellent for the vinegar manufacture; it should, however, be carefully collected, and the first portions of the showers which wash the roof, should be rejected. These precautions should especially be observed with lead, copper, or zinc roofs, as these metals would bring poisonous substances into the vinegar. In coal burning localities, the first rain water is very dark from soot. Rain should always be filtered through charcoal to remove organic matter which it contains, and which is capable of putrefaction. A filter can be made with very little trouble, using a large wooden vessel, but the following one proposed by Dr. Otto leaves nothing to be desired. A cubical vessel is made by bolting together slabs of sandstone, the same being rendered water-tight by hydraulic cement. A hole at the bottom contains a suitable faucet.

The filtering strata are: First, on the bottom, an eight inch layer of stones or fragments of brick of the size of an egg, and arranged carefully to permit the flow of water through all parts of the layer.

Next a layer of coarse gravel; then coarse sand; next layers of charcoal in strata of different fineness; above this a layer of coarse sand, and above all a good layer of fine sand. The water must enter the filter in such a manner as not to disturb the upper layer of sand. Of course, all of these materials must be well washed to remove *all* muddy or fine particles, and when the filter is first set in operation, the water flowing through it must be rejected as long as it continues cloudy.

When the filter, after long usage, ceases to clarify the water, the top layer of sand must be renewed by fresh sand or by washing. A filter of this kind will very soon pay for itself in vinegar factories where the water on hand is not perfectly clear, and free from organic matter.

IV. *The Factory Building.*—The minor details of the building will, of course, vary with taste, convenience, experience, and the particular kind of vinegar manufactured; but some general details are observed by all. By the slow process, as will be seen hereafter, a much larger building is needed. By both processes the preservation of an equable temperature is desirable, and consequently a southern exposure is generally desirable for the building. The ventilation of the vinegar-room should be under complete control, by properly placed air-flues and registers. The violence of the summer's heat may be prevented by ventilation and by employing cold water for making the wash with the judicious use of ice. Double walls to the building, double windows and a vestibule to the door, are very useful, especially in the slow process. In winter-time the heating apparatus should be fed with fuel from without to avoid a two frequent opening of the vinegar-room. To save fuel, the height of this room should be no greater than conveniently to contain the apparatus. The floor should be dry and tight; preferably of brick or stone. The number and size of the windows should be as small as convenience will allow. The walls and ceiling should be furnished with a good coating of plaster of paris, as the acid vapors of the manufacture will inevitably and speedily attack a lime coat, disintegrating it, and causing it to crumble off. For the same reason all iron and other metallic work must be kept well painted, or coated with asphaltum varnish. Vinegar, should never, during its manufacture, be brought in contact with any metal especially with copper, lead or zinc. The factory should contain conveniences for warming the water and vinegar mixture. When its capacity will permit, steam effects the best purpose in this respect.

CHAPTER II.

THE SLOW PROCESS.

The difference between the slow and quick vinegar processes may be illustrated by the following example: Ten gallons of 80 per cent. alcohol added to 130 gallons of water, give a wash of 6 per cent. alcoholic strength. If 40 gallons of vinegar be added, and the wash exposed to air of the right temperature, 180 gallons or $4\frac{1}{2}$ barrels of vinegar, containing from 4·6 to 4·8 per cent. of anhydrous acetic acid will result. By the slow process, this vinegar could be made by placing the mixture in vessels, and maintaining the temperature between 74° and 86° Fah. The time required for the manufacture would vary with the temperature at which the vinegar-room was kept, and with the size of the acidifying casks. If the temperature was moderate, and the casks of half a barrel capacity, the requisite time would be about 16 weeks. From which it will be seen that for the manufacture of a large quantity of vinegar yearly, an extensive factory building would be requisite, much capital must be invested in the fixtures, and four months interest upon a large amount of material would have to be accounted for.

By the quick process, the same vinegar mixture

would be run through large tubs, called "*generators*," or "*graduators*," filled with beach shavings, arranged so that a current of warm air is brought in contact with the wash thus diffused over a large surface. The $4\frac{1}{2}$ barrels of vinegar would be made in $4\frac{1}{2}$ days instead of as many months required by the slow process. Not only is time saved by the quick method, but fuel also; for the heat generated by the oxydation of the 10 gallons of alcohol instead of being lost by extension over a period of months, is available when condensed into as many days, nor has the room to be heated for so long a period. For the above proportions, the manufacturers by the quick process would require a small room with two generators each, capable of transmitting a barrel of wash in 24 hours, for which an insignificant capital would be needed. Let us now consider the slow process as practised upon a small scale, and as in large manufactories.

I. *The Household Manufacture.*—A great deal of vinegar is thus made from wine, cider, and in some places from alcohol. In some parts of Europe the vinegar cask descends through the same family for several generations. The remainder of their daily wine is poured into the cask and vinegar drawn therefrom as required. In our country cider mixed with water is poured into a barrel painted black, and the bung-hole left open. Its color enables all the heat of the sun to be made available in cool seasons. The best cider contains about 9 per cent. by volume of alcohol, which is equivalent to a per centage by weight of $7\frac{1}{2}$ anhydrous acetic acid; as this is a stronger vinegar than generally needed, water must be added to the cider. A 6 per cent. alcohol will yield a vinegar of from 4·6 to 4·8 per cent. acid strength. The addition of half a gallon of water to every *gallon* of the best cider, (in an article containing 9 per cent. alcohol,) will reduce its alcoholic strength to 6 per cent. Because by the rule given on page 155, $\frac{9}{6}=1\frac{1}{2}$, that is one gallon of 9 per cent. alcohol yields $1\frac{1}{2}$ gallons 6 per cent. alcohol.

The following household vinegar method is to be recommended as simple, expedient, and furnishing a constant supply of vinegar with scarcely any trouble, and at trifling cost:

Two barrels are procured, one for making, the other for storing the vinegar. Those from which *good* vinegar has just been drawn are preferable. The storage barrel is kept always in the cellar, the generating one in the cellar, or house, according to the season. In this latter barrel a small hole is bored, for the circulation of air, at the top of one of its heads. The barrels lie on their side, and contain each a wooden faucet. Of course their capacity is regulated by the yearly demand of vinegar.

We will suppose that the generator, filled to the level of the ventilating hole, contains 10 gallons, the manufacture will then be carried on in the following manner. Seven gallons of good vinegar are placed in it, and three gallons of a warm alcoholic mixture, made in the following manner, are added. If common whisky (50 per cent.) be employed, have a small measure of 3 pints, and a large one (a bucket) of 3 gallons. If 86 per cent. spirits are used, let the small measure be for 2 pints. Put a small measure full of the spirits in the large measure; fill quickly to the mark with boiling water, and pour by a funnel into the generator. Every two or three weeks, three gallons of vinegar are withdrawn from the generator, added to the storage barrel, and three gallons of alcoholic mixture are placed in the generating barrel as before.

Another method of working the casks consists in half filling the generator with vinegar, and adding every week so much of the alcoholic mixture that it fills the barrel in from 8 to 16 weeks, according to the season. Half the vinegar is then added to the storage cask, and the process recommenced in the generator. The warmer the season the more rapid may be the manufacture.

II. *The Factory Process.*—The process which I have just described for the household does not differ very materially from that of the factory. In the latter greater care is taken of the temperature; and as the quantity of material operated upon is large, and contained in many casks, each cask is carefully watched, not only to remove the vinegar when made, but to conduct the process with regularity, and to prevent putrefaction arising in a cask and spreading to its neighbors. The most celebrated locality for the slow manufacture is at Orleans, France, where wine vinegar is made. The following description of their method* will be instructive, and affords a very good example of the principles involved in the slow process.

The wine is first clarified by standing for some time in large vessels filled with beech shavings, upon which the lees deposit. It is then drawn off gently from the bottom, ready to be converted into vinegar. The fermenting casks are of 75 gallons capacity, and are arranged upon their sides in four tiers. They are pierced at the upper part of their front heads with two holes—a large one for the funnel which introduces the wine, and for withdrawing the vinegar, and a smaller one for the access of air. When the process is commenced the casks are one-third filled with the very best vinegar, which is the "*mother*," of all subsequently made. The casks last for twenty-five years, but must every ten years be cleansed and started afresh, on account of the accumulation of tartar and lees, which retard the acetification. The temperature of the apartment is maintained between 75°–77° Fah., by means of wood fires in cast-iron stoves. As the top rank of casks is in the warmest position, "lazy" casks, as those are called in which the fermenta-

* From Dr. Ure.

tion is too slow, are removed to that rank. When the manufacture is in full operation, a charge of 10½ quarts (10 litres) of clarified red or white wine is added to each cask every 8 days. After four such charges, 42 quarts of vinegar are withdrawn, and the operation recommences. Matters are so arranged that the casks are never more than two-thirds full. The workman's skill consists in managing the temperature, and watching the casks, so that the fermentation shall proceed with uniformity. Before the vinegar is withdrawn the following empirical test is applied to discover whether the fermentation is complete. A white stick, curved at one end, is plunged into the vinegar through the large orifice of the cask, and withdrawn horizontally. If covered with a thick *white* froth, the acetification is perfected; if, on the other hand, the froth is *red* and *pearly*, the cask is lazy, and must be brought to a more vigorous action by increased temperature and by the addition of more wine.

Let us now consider a factory constructed upon a similar principle, for the preparation of vinegar from whisky or other spirits. Let the casks be arranged in tiers with a hole in each for ventilation and introduction of the wash, and a wooden faucet for withdrawing the vinegar. The heating apparatus may be either stoves, a hot-air furnace, or a similar arrangement to that employed for heating green-houses. In fine, let due attention be given to the methods of maintaining an equable temperature detailed in Part II., Chapter 1st. There should be appliances for heating water and also vinegar. The latter may be effected in hot water baths containing large glass bottles, vitriol carboys, demijohns, &c.

If the casks are new, they must be prepared by repeatedly soaking inside with hot water, and finally several times with warm and strong vinegar, to which a little alcohol has been added. The object being to remove extractive matter from the pores, and fill them with the vinegar ferment.

After everything is arranged we must decide upon the strength of the resulting vinegar. Table II., page 204, will give the necessary information. Column 1 denotes the degrees of Tralles alcoholometer the wash should indicate (in other words its per centage by volume of absolute alcohol) for a corresponding strength of the vinegar given in column 7th.

On page 154 will be found the rule for making a mixture of required per centage from strong alcohol and water. Suppose it were required to make a vinegar containing about $4\frac{1}{2}$ per cent. of anhydrous acetic acid. Theoretically by Table II., page 204, the wash should contain a little over 5 per cent. absolute alcohol; but on account of the loss by evaporation of alcohol we would have to take a wash of 6 per cent. If we employed for making this wash 80 per cent. spirits, then, since $\frac{80}{6} = 13\frac{3}{10}$ we would have to dilute the spirits so that every gallon becomes $13\frac{3}{10}$ gallons. In other words, to 100 gallons of 80 per cent. spirits we add 1230 gallons of water, equal 1330 gallons of mixture, which, after the addition of 300 gallons of vinegar to act as a ferment becomes 1630 gallons of wash. A portion of the water must be taken sufficiently hot to give a temperature of from 90°–100° Fah. to the wash. The resulting wash is placed in the fermenting casks to fill each one two-thirds, and the temperature of the apartment, observed by thermometers placed in different parts of it must be kept between 75° and 100° Fah. At the minimum temperature less fuel is required, but the time needed for the acetification is extended, and consequently more casks and a larger apartment are needed to make the same amount of vinegar. With the maximum temperature the reverse takes place.

Several days after the addition of the wash, the acetification begins, and is indicated by a temperature in the casks elevated slightly above that of the apartment. A piece of slate laid over the large hole in the cask, to prevent too great evaporation and consequent cooling, is bedewed with moisture, and the pleasant odor of vinegar is perceived in the room. As long as these indications continue, everything is going on well; but every cask must be examined by itself to at once restore the action in any "lazy" ones, lest putrefaction or mouldiness take place and spread to the neighboring casks. When this misfortune occurs, the bad casks are at once removed from the apartment, *their contents thrown away*, and the casks scoured well with brushes and water, and placed in the sun. After they are dry, they may be soaked in hot vinegar, and brought into action again. If only "lazy," they are excited by withdrawing a portion of their contents, which is warmed in glass bottles, and with the addition of a little spirits and vinegar is restored to the casks.

Sometimes too cool a position, or a constant draught of air will bring a cask out of action. The remedy is, removal after the acetification is restored to a warmer position, (an upper tier,) or by covering with a non-conductor, as, strong paper pasted over it.

If the staves of the casks, especially of the lower tier, are thick, the disadvantages of a cool position are in a great measure overcome.

After a lapse of time depending upon the temperature, which is kept a little more elevated towards the close, the acetification is complete. Otto gives the following as the time generally required.

For temperature, Fah.	Weeks required.
100° — 95°	4 to 6.
95 — 86	6 to 10.
86 — 80	10 to 12.
80 — 7?	12 to 20.
below 73	8 to 10 months.

The close of acetification is indicated by the diminution of the strong vinegar smell in the

THE FACTORY PROCESS. 229

apartment, by the absence of vapor condensing upon the slate covers of the holes, and by the temperature of the inside of the cask becoming equal to that of the room. In the process just described, either Otto's acetometer, (page 189) or Balling's vinegar tester, (page 202) will perform excellent service to indicate the march and completion of the acetification. The indications of the instruments, together with the date of the experiments should be chalked upon the heads of the respective casks. By this means, the condition of any cask in the establishment may be known at a glance.

As soon as the acetification in any one cask is perfected, the vinegar must at once be withdrawn, barreled and removed to a cooler place than the vinegar room, in which its tendency to spoil in the heated atmosphere is very great. The slimy deposit called "mothers" is removed, and the vinegar with which it is imbued, employed in part for the next acetification. If the sediment from each barrel be placed in a cask, the clear vinegar may be drawn off after the deposition of the mothers. It is well, before barreling the vinegar, to suffer it to stand for a short time, in a cool place, in a vessel filled with beech shavings, which clarify it. When stored, a pint of spirits should be added to each barrel.

The slow process, thus described, may be modified in various ways:

1. Thus, instead of bringing the fermentation

to completion in all of the casks at about the same time, they may be divided into 3 or 4 groups, so that $\frac{1}{3}$ or $\frac{1}{4}$ of the whole quantity of vinegar may be withdrawn and stored at intervals of $\frac{1}{3}$ or $\frac{1}{4}$ the time required for the acetification of the whole quantity. This modification has the advantage of a greater distribution of the work; necessity for a smaller quantity of vinegar stored for sale; and the presence of casks in full action, emitting strongly acetic vapors, which is of advantage in keeping up the fermentation in casks just going into operation. The disadvantages consist of a greater need for entering and leaving the vinegar room, involving loss of its heat, and requiring in consequence, greater attention to its fires. In addition to this, the heat cannot be increased towards the close of acetification, which is useful in shortening the time required for the manufacture.

2. Another modification consists in always keeping a large quantity of vinegar in the fermenting casks, and at short intervals withdrawing small quantities of vinegar, which are replaced by fresh wash. This saves time, as acetification is more rapid in the presence of large bodies of vinegar. It involves loss of heat by a need for a too frequent entering the vinegar room. It involves, also, a loss of interest upon the value of the large quantity of vinegar kept in the fermenting casks. The intervals at which vinegar may be withdrawn are closer in propor-

tion to the heat of the apartment, which bears a ratio to the amount of fuel consumed.

By this method, only $\frac{1}{5}$ of the vinegar is removed at one time from each cask; in other words, at intervals of from one to two weeks, according to temperature, 1 gallon of vinegar is withdrawn from every 5 gallons in the fermenting casks, and in its stead a gallon of wash is added. In a large factory, the latter process requires a large number of barrels of vinegar to commence operations. This vinegar must be either purchased or made gradually in the fermenting casks, not withdrawing any until the casks are sufficiently full. The advantage consists in the need for a smaller number of fermenting casks, than by the method first described. Dr. Otto gives, in his treatise on vinegar, the following calculation for the number of fermenting casks required for the slow process:

Suppose that it be required to furnish a barrel of vinegar per day excluding Sundays, which would equal 312 forty-gallon barrels per year, the fermenting casks would have a capacity of $\frac{1}{2}$ a barrel, and since they are not *filled* with wash, and on account of unavoidable loss, we may allow 4 such casks to each barrel of vinegar made. We, of course, do not account as manufactured vinegar what is added to make the wash, as a like quantity must be added in the subsequent wash. From every four fermenting casks we may sell one barrel of vinegar; hence, six barrels of vinegar will require 24 such casks. If the vinegar room be so heated that the operation is completed in four weeks, we will have to draw off 24 barrels of vinegar, to do which 96 fermenting casks will be required. If, however, a lower temperature be maintained in the apartment, say to complete the process in sixteen weeks, 4 times 96 = 384 fermenting casks will be required. In the latter case, the expense of fuel is lessened, but that of the fermenting casks is increased. Besides, a larger apartment will be requisite, which will involve a higher rent and greater expense for fuel in heating it.

If the process be modified, as described, so that a large body of vinegar is always kept in the fermenting casks, their number may, as before stated, be proportionally decreased.

This calculation affords the very best illustration of the superiority of the modern quick process, over the ancient slow method. To make one barrel per day by the quick process, a small room and two "generators" are the sole requisites.

CHAPTER III.

THE QUICK PROCESS.

I. BOERHAVE'S METHOD.—After the vinegar process was thoroughly comprehended, the discovery of the quick method was to be expected; for, since the action of the air at elevated temperatures performs the acetification of alcohol, what was more natural than to construct vessels by means of which the alcoholic liquid could be brought in contact with a larger proportion of air in a given time than by the old method. The modern process is not much more than a score of years old. As, proverbially, coming events are frequently foreshadowed, so it has been with the vinegar process; for, in the 17th century, Boerhave introduced an improvement upon the vinegar manufacture of his day, which includes the principles of the modern process, and which has obtained considerable employment, especially in France.

This improvement was as follows: Two roomy casks of equal size are taken and placed upright in the vinegar room. The superior heads are removed, and the bungs driven in. They are then filled with the stems resulting from depriving bunches of grapes of their fruit. One cask is filled, the other half filled with wine, to which ¼ vinegar has been added. Every 12 or 24 hours half the liquid of the full cask is poured over the stems in the half empty cask, which it of course fills. Thus each cask is alternately full and half full. The alcoholic mixture spreads itself over the grape stems, which rise in a porous heap into the air of the half empty casks; these stems soon become coated with "mothers," and the acetification takes places rapidly. This process was subsequently so improved that the mixture was poured every 3 or 4 hours. By this means a vinegar was obtained in 14 days that would by the old process have required months for its manufacture. Boerhave's method differs from the modern practice, only that in the latter a larger porous mass is obtained in the generators, and care is taken for a circulation of air.

II. DŒBEREINER'S METHOD.—Dœbereiner sought to introduce a method of the vinegar manufacture, in which the action of finely divided metallic platinum was substituted for a ferment. The process obtained a trifling application here and there; but could not become of universal use on account of the very large capital which must lie dead in the platinum. The metal is not deteriorated in the process. I will describe briefly the process from the light it throws upon the quick process, showing that the vinegar "*ferment*" is only one of the *instruments* of acetification, and not a sine qua non. Spongy platinum, or platinum black, which is the efficient agent of this operation, is prepared in the following manner. Grains of the native metal, or cuttings of the pure metal are boiled in glass or porcelain vessels with aqua regia, (3 parts hydrochloric acid, and 1 part nitric.) The residue is treated in the same manner until complete solution of the metal, which requires for every ounce of platinum from 10 to 15 ounces of aqua regia, according to the size of the grains. This solution may be effected with less cost by first melting one part of the platinum with from 2 to 3 times its weight of zinc, and granulating the alloy, which is then acted upon by diluted oil of vitriol, which dissolves the zinc and leaves the platinum in a state of fine division. This powder is further purified by boiling in a little nitric acid, when it becomes readily soluble in aqua regia.

When a clear concentrated solution of platinum is obtained, add to it solution of sal ammoniac, which throws down, as a yellow crystaline precipitate, the double chloride of platinum and ammonium. These crystals are thrown upon a filter, and washed with a little water. When heated, everything is volatilized but metallic platinum, which is in a very fine state of division, called platinum sponge or black.

This substance, from a property not yet completely understood, causes oxygen to unite with gases having an affinity for it. Thus hydrogen and oxygen may remain mixed forever without combination taking place. If a piece of spongy platinum be placed in the mixture, the gases at once unite with explosion.

In the same way alcohol vapor and air will under the influence of spongy platinum, unite so as to form vinegar.

Dœbereiner's "lamp without flame" illustrates this action. A thin platinum wire, holding a ball of spongy platinum, is placed above the wick of a spirit lamp, that the ball of sponge may be made red hot by the flame of the lamp. The flame is then dexterously blown out so as not to cool the platinum, which continues to glow heated by the oxydation of the alcohol vapors rising from the wick, and which thus burn without flame, giving rise to *aldehyde*, recognized by its peculiar smell. The platinum causes the oxydation of the alcohol so rapidly that, sufficient air to perform the complete transformation to vinegar cannot be brought in contact with it, and aldehyde results instead.

The perfect change may be effected in the following manner.

Fill a saucer with 10 per cent. alcohol, and place the platinum black in a little vessel, supported by a glass triangle resting on the saucer. Cover with a bell glass, standing in the saucer, and place in a light place, (in the sunshine if convenient,) where the temperature is from 68° to 86° Fah. The alcohol vapors will rise, become mingled with air, and then converted into acetic acid by action of the spongy platinum. The vinegar will condense upon the sides of the bell, and trickle down into the saucer, which will at the close of the operation be filled with vinegar. It may be ascertained by a thermometer that the temperature rises in the bell glass. In order that the air of the bell glass be constantly renewed, the bell must be tubulated and closed loosely by covering the tubulus with a plate of glass. 1000 cubic inches of air can oxydize 110 grains of absolute alcohol, giving rise to 122 grains of acetic acid, and 64½ grains of water.

Dr. Ure estimates that with a box of 12 cubic feet capacity, and from 7 to 8 ounces of spongy platinum, 1 pound daily of alcohol can be converted into acetic acid, and that with from 20 to 30 pounds of platinum we may obtain 300 pounds of vinegar from the same amount of spirits. The costliness of the operation, as illustrated by this example of Dr. Ure, lies in the capital buried in the platinum, and explains readily why Dœbereiner's method has found no favor with the public, especially when we consider the small outlay for the apparatus of the quick process.

III. THE MODERN PROCESS.—A concise and full description of the quick method of the vinegar manufacture, would fall naturally into the following subdivisions:

I. The apparatus.
II. The details of the operation.
III. The experience of the best factories, and

IV. Some remarks upon the question of the expediency of large factories.

I. and II. will be considered in the present, and III. and IV. in the following chapter.

I. THE APPARATUS.

The requirements for this process consist of a large room of convenient size, capable of being maintained naturally or artificially at an equable temperature, between the limits of 74°—86° Fah.

The water should be accessible, good, and under the circumstances alluded to on a former page, filtered.

The storage accommodations for the manufactured article should be cool, dry, and free from mouldiness.

In order to ascertain the exact cost of his product, the vinegar-maker may possess—

1. Two Tralles alcoholometers—one for ordinary, the other for extremely dilute solutions of alcohol.
2. Otto's Acetometer and its appurtenances.
3. Balling's Vinegar Tester.

The factory should contain at least two "generators," filled with beech-shavings, and for *each* of these, two small tubs of the capacity of a barrel, one for containing the alcoholic mixture, the other for the acidified product. There should also be, according to some methods of performing the operation, a larger mixing tub, and a gallon measure for the preparation of the "wash."

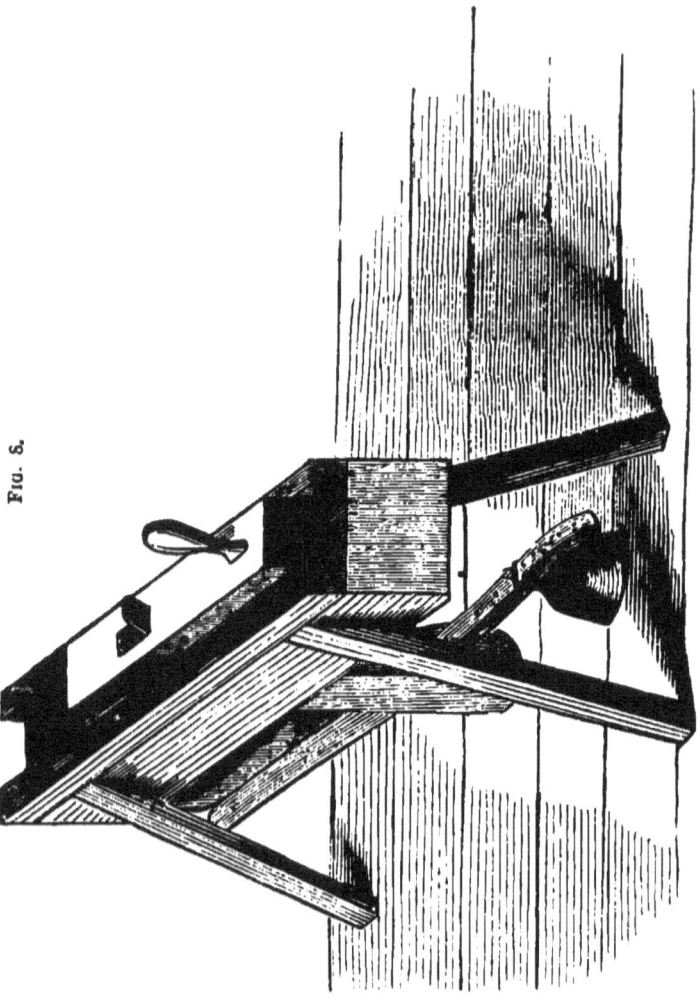

Fig. 6.

Arrangements for heating water to prepare the wash are necessary, and by some methods, conveniences for heating in glass bottles wash that that has passed a generator, and which contains, in consequence acetic acid.

The Shavings.—If beech-wood shavings of the required shape and quality cannot be purchased, the following tool for their manufacture will be required. (See Fig. 8.).

It consists of a heavy plane, which by reason of rebates attached to its sides, is capable of only a backward and forward motion in a frame. The floor of the frame is perforated with a longitudinal aperture, in which a beech-wood board of a foot in length, 1 inch in thickness, and 6 to 8 inches in breadth, may slip and present its edge to the plane iron, which cuts from it shavings 1 inch broad by a foot long. By means of a lever and weight these beech-boards are pressed upward with the desired force to enable the plane to cut, and the plane is never drawn far enough back for the beech-boards to escape upwards from the frame. Closely curled shavings are required on account of their strength, and porousness; the close curl prevents the lower layers in the generators from being crushed by the weight of the superior layers.

To obtain a shaving of any required spiral, the plane is doubly ironed. As the sharp iron cuts the wood the shaving is pushed over by the blunt iron, (to a degree proportional to the proximity of the blunt iron to the edge of the sharp iron,) forming a close curl. With beech boards of the above dimensions six shavings have the area of a square foot.

The generators are tubs slightly conical, with the upper diameter the larger. They are made of oak or white pine, and may be from 6 to 12 feet in height, and from 3 to 4 feet in diameter. At from 8 to 14 inches above the bottom of the generator, 6 or 8 half-inch air-holes are bored, at a slightly descending angle, so that the wash trickling down the inside of the tub may not escape through them. At the distance of a couple of inches above the air-holes, a false bottom, pierced with $\frac{3}{4}$ inch or inch holes, is placed to sustain the shavings. From 6 to 8 inches below the top of the generators, a hoop or else three isolated cleets are fastened with wooden nails, for the purpose of supporting a horizontal partition pierced with small holes, through which the wash rains down upon the shavings. The holes are of a line in diameter, placed at regular distances apart of $1\frac{1}{2}$ inches. They should be bored, and be reamed out with a red hot iron to prevent their swelling shut by the action of the hot vinegar mixture. The crevices between this partition and the walls of the generators should be carefully caulked with oakum. Some prefer to a partition a shallow tub, fitting in the top of the generator, and with the bottom pierced. In both cases, pieces of twine or straw

THE APPARATUS.

are drawn through the holes, along which the alcoholic wash trickles to fall in drops upon the shavings. Six inches below the sieve a hole is bored at a descending angle, for the reception of a thermometer, of which the scale must be visible from 75° Fah. upward.

Besides the small holes in the sieve, four larger ones are bored at equal distances apart. They provide for the insertion of glass or wooden tubes of ¾ inch bore, and which reach to within an inch of the cover of the generator. These are chimneys by which the draught of air escapes from the generator. They must fit water-tight.

FIG. 9.

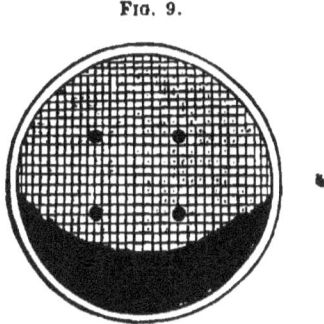

Figure 9 represents the sieve-tub seen from above. The small holes are bored at the angles of the squares. The sieve-tub or partition must be made so as not to warp, and must be placed in a perfectly horizontal position. A cover must fit well upon the generator, and be caulked, if not air-tight. In the middle a hole is made for the introduction of the wash. It must be capable of being closed to a required degree, to aid in regulating the quantity of air passing through the generator. It is a fixed principle in the operation of a vinegar generator to suffer no more air to traverse it than what is just sufficient to effect the acetification, as otherwise more alcohol is lost by evaporation. The generator contains, near the bottom, a large wooden faucet for withdrawing its liquid contents, and also a small hole for the reception of a goose-neck tube of glass or hard vulcanized rubber. The upper curve of this tube must be below the air-holes; its use will be described directly.

Fig. 10.

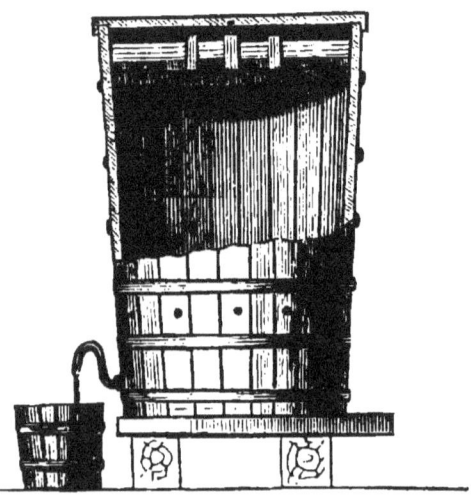

Figure 10 illustrates a generator as just described. Otto gives a couple of modifications of the generator as follows:

In place of the lower air-holes, the *bottom* of the

THE APPARATUS. 245

generator is pierced in the centre and furnished with one air-tube shaped something like a nine-pin; (see fig. 11,) its head reaching as high as the

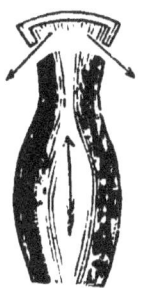

FIG. 11.

lower air-holes of figure 10. Below the closed dome-shaped top, it is pierced with several holes slanting upward and inward. The shape of the tube effectually prevents the escape of any drop of vinegar which may fall upon it, while it furnishes air to the very centre of the shavings, diffusing it more equally than in the generator, figure 10. If it rises above the bend of the goose-neck no vinegar can escape by it to the floor.

Another practice consists in retaining the lower air-holes, but fitting in them air-tight glass tubes at a descending angle to prevent escape of vinegar, and reaching nearly to the centre of the tub.

I have seen generators furnished with wooden tubes of this description, but which entered a longitudinal box pierced with holes, and extending through the body of shavings from the bottom to near the top. This arrangement is very

21*

bad, as the tendency of the air is to draw through the box which acts like a chimney, and is a shorter and freer route than the interstices of the shavings.

Another modification of the generator relates to the goose-neck. The object of this tube is to keep constantly in the generator a large body of warm acidified wash, which stands at the level of the upper bend of the tube. This utilizes the heat which the wash has acquired by the acetification, and thereby spares fuel.

Figure 12 illustrates the lower portion of a generator in which is a faucet of which the portion

FIG. 12.

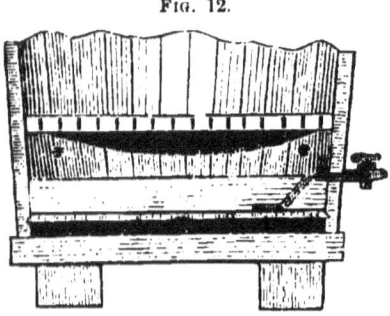

prolonged into the tub answers the object of the goose neck. This faucet is left open and the liquid flows from the generator when its level corresponds to the delivery of the faucet. A second faucet near the bottom of the generator is needed to draw off when necessary, all of the vinegar. Besides the advantage of saving heat, the vinegar drawn from the bottom of the tub, as by the action of the goose neck or faucet of figure

12, is of stronger quality, for several reasons, which will be given in the proper place.

The generators should, of course, be so placed as to permit their liquid contents to flow into the necessary barrels or other vessels in the course of the manufacture. They should also have space above them to facilitate delivering the wash to the sieve tub, whether by tubs or by hand. More space than is just sufficient is injurious, involving a waste of heat; a light gallery or merely a ladder enables the tops of the generators to be reached. It is not generally a good plan, at least in winter, to have tubs delivering the wash in an upper story; as the upper portion of a room is the warmest, heat may be saved by placing the tubs for delivering the alcoholic mixture near the ceiling of the vinegar room.

The wash may be elevated by hand or by pumps of wood, glass, or hard vulcanized rubber.

II. DETAILS OF THE OPERATION.

When everything is ready, the fermentation of the generators is brought about, as follows:—For a day or more, the vinegar room is kept at a temperature between 75°—80°, in order that its walls, the generators, etc., acquire the proper temperature for the acetic transformation.

The beech-wood shavings must then be boiled in good strong vinegar and placed in the generators. This heating with vinegar may be effected in a wooden tub with a pipe delivering steam into it. I have found that as good a way as any is to boil the shavings in 20 gallons of vinegar in a perfectly clean cast-iron cauldron, placing them at once in the new generators and rejecting the first vinegar made. While one batch of shavings is heating in the vinegar, the preceding batch which was raked out of the cauldron with a pitchfork and carried to the generator in a basket, is placed carefully in the situation it is to occupy, settling it with a rammer made by fastening a barrel head to a pole. The ramming is to be gentle so as not to break the shavings, which are to be arranged as a uniformly porous mass, and to leave the air holes free, remembering that the generator is a stove in which the shavings are, at the same time, fireplace and chimney. When the layers of shavings have reached to within 6 or 8 inches of the sieve bottom, the same, furnished with its ventilating tubes and the twine in its holes, is placed in position, and the cover fastened down securely.

The shavings at the thermometer hole and at the lower ventilating holes are then loosened by means of a stick thrust therein. A wash is now prepared which contains $\frac{1}{3}$ vinegar and $\frac{2}{3}$ of a 3 per cent. solution of alcohol; this, heated to 75° —80°, is gradually poured through the hole in the cover of the generator, at the rate of one barrel in the lapse of 24 hours. At the expiration of this time, warm the resulting vinegar if necessary, and having added enough alcohol to make the whole quantity taken thus far, of 5 per cent. alcoholic strength, pour this through the generators as before. Repeat this operation on the third and even on the fourth day if it be necessary. Investigate the temperature of the air escaping from the generator, and when it *exceeds* that of the wash which is running, it is a sign that the acetification has commenced. When it rises to a point between 98°—104°, the generators are in a proper condition to commence the regular business of the manufacture; the fermentation has been properly established. We then daily pour through generator, No. 1, a wash consisting of a certain quantity of spirits, vinegar and water, heated to a temperature between 75°—80° Fah.; and through No. 2 the wash which has passed through No. 1, to which has been added more spirits. We draw manufactured vinegar daily from generator, No. 2. The vinegar resulting from setting the generators in action, though not prejudicial to health, is of inferior quality and bad flavor from extractive matter from the shavings and tubs and from the iron cauldron. It may be added in very small quantity to the subsequent vinegar, if it be not thrown away.

I have thus sketched, in a few words, the result of my own experience as the most convenient way of commencing operations with new generators. Many modifications may be made: instead of shavings, chips of beech wood, corn cobs, charcoal and several other porous bodies have been employed. In Germany, especially, the use of charcoal has been followed with great success; the coal is broken in pieces of the size of a walnut, sifted from dust, washed and dried. When saturated with vinegar, it acts precisely like beech shavings the pores of the coal absorb from five to six times as much vinegar as beech shavings, which brings a greater amount of ferment in contact with the wash. The air *surface* is not increased by this porosity, the pores being filled with liquid.

The following is another and better method for bringing generators into action, but requiring more time and labor: the generators, shavings, and all the vessels employed are, if new, well soaked in warm water, which is renewed several times if necessary for the purpose of dissolving soluble matter. The shavings may be soaked in the generators, after which, they must be spread out in some convenient place to dry quickly. After the generators are also dry, the shavings are packed in them with the proper precautions, and every thing is prepared for the manufacture. The generators may then be acidified in the following manner: the temperature of the apartment having been kept for a short time between 75°—80° Fah. Hot and strong vinegar is repeatedly poured upon the sieve bottom in such a manner that it may drop evenly over the stratum of shavings. This vinegar may be heated in an iron cauldron, perfectly scoured; but the best mode of warming is by large bottles, heated in a hot water apparatus, since salts of iron, entering the vinegar, would form ink, with the astringent substance left in the material of the generator. By this method also, the vinegar extracts what soluble matter from the shavings and generators is left undissolved by the water employed in the soaking, and in consequence the first vinegar made is of inferior quality. As the generators begin to become well saturated with vinegar, a little alcohol may be added to form a wash; increasing gradually the quantity of alcohol added, until we thus pass imperceptibly into the vinegar manufacture. At this juncture, it is well to close the opening in the cover of the generator for 12 or 24 hours; if, at the expiration of this time the thermometer indicates a temperature of 95° to 100° when placed at the opening in the cover, the generators are in a proper fermentative condition for proceeding with the manufacture.

The wash may be poured in two ways, either at intervals, (every hour,) in definite quantities, according to the size of the generator, and interrupting the process at night; or it may flow uninterruptedly in a measured stream. The relative advantages of these plans will be discussed upon a future page. To make a good, strong vinegar, three generators should be used. The wash passes through each in succession, with the addition of alcohol each time; by this means a 6 per cent. alcoholic wash yields a vinegar containing from 4·6 to 4·8 per cent. of anhydrous acetic acid.

With a 10 per cent. alcohol passing the shavings four times, a stronger vinegar may be made, containing about 8 per cent. anhydrous (= $9\frac{1}{2}$ hydrated) acetic acid.

Practical example of the manufacture.—The following example will illustrate practically the manufacture of vinegar by the quick process. We will suppose three generators of from 10 to 12 feet high to be employed, and that a wash, receiving at each time alcohol, passes each generator in succession, leaving the third in a state of manufactured vinegar.

As it is immaterial for the example whether the wash runs uninterruptedly or is poured at intervals, we will imagine the former case, and that it runs at the rate of $2\frac{1}{2}$ gallons per hour, that is, yielding 35 gallons every 14 hours. Let the object be to make a vinegar of 4·6—4·8 per cent. acid strength, which, as we know, will require a mixture containing 6 per cent. of absolute alcohol. Six mixing tubs, of 40 gallons (a vinegar barrel) capacity, will be required. This number will enable the work to proceed without delay in case the wash should flow more slowly in any of the generators.

Let us first consider the proportions of the wash. It should contain $\frac{1}{9}$ of manufactured vinegar, and $\frac{8}{9}$ of a 6 per cent. alcoholic wash. By the mixture rule, 1 gallon of 80 per cent. alcohol added to $12\frac{2}{10}$ gallons of water, yields $13\frac{2}{10}$ gallons of 6 per cent. alcohol. To make

DETAILS OF THE OPERATION. 253

31 gallons of such weak spirits, we require $28\tfrac{1}{2}$ gallons of water, and $2\tfrac{1}{2}$ gallons of 80 per cent. alcohol; because $\tfrac{1}{13}$ of 31 gallons $= 2\tfrac{4}{10}$, or nearly $2\tfrac{1}{2}$ gallons. To these 31 gallons 4 gallons of manufactured vinegar must be added, yielding 35 gallons of wash, which has to run during 14 hours. This wash then, to repeat, consists of

Water,	$28\tfrac{1}{2}$ gallons.	
Vinegar,	4 "	
80 per cent. alcohol,	$2\tfrac{1}{2}$ "	$= 10$ qts.
	35 "	

The result of three runnings, is vinegar of 4·6 to 4·8 per cent. acid strength. The $2\tfrac{1}{2}$ gallons of 80% spirits, are added in three *unequal* portions one before each running. A moment's reflection will show that the division must be unequal, for not only is a dilute solution acidified more rapidly, but there is greater loss of alcohol by evaporation in a stronger solution dropping through the air current in the generator. Some unconverted alcohol always passes the shavings. If, therefore, we divided the 10 quarts of spirits, into three portions of $3\tfrac{1}{3}$ quarts each, adding one to each wash, the first wash would contain $3\tfrac{1}{3}$ quarts of spirits; the second by means of unconverted alcohol passing the first generator, more than $3\tfrac{1}{3}$, and the last wash still more, involving in the latter washes, loss and inconvenience from the causes stated. The best division would be

22

254   VINEGAR MANUFACTURE.

```
To first wash    .    .    .    6 quarts 80% spirits.
To second wash   .    .       2½    "         "
To third wash    .    .    .  1½    "         "
                              ———
                               10   "         "
```

The mode of operating then would be to run through the first generator a mixture of 28½ gallons of water, 4 of vinegar, and 6 quarts of 80% spirits. After it had passed, to add 2½ quarts of spirits, and let it traverse the 2d generator; after which, with the addition of 1½ quarts of spirits, it would leave the third generator as 35 gallons of manufactured vinegar. Since four gallons of this was added vinegar, the result is 31 gallons of vinegar made. After every working day of 14 hours, so much vinegar can be stored. It is equal to $3\frac{1}{10}$ forty gallon barrels every four days.

The following scheme will illustrate the mode of operating the three generators and six mixing tubs.

```
        1        3        5
        A        B        C
        2        4        6
```

The letters denote the generators, and the numbers their respective mixing tubs. The generators are furnished with goose-neck tubes to keep always a body of warm vinegar in them, consequently when in full operation, for every drop of wash that falls from 1, 3, and 5, a corresponding drop leaves the goose-neck of A, B, and C, to fall into 2, 4, and 6. The mixing tubs have each a mark to indicate the level of 35 gallons. At the commencement of a day's work, 1, 3, and 5 are empty, and 2, 4, and 6 contain about 35 gallons each of vinegar of varying strength, increasing from 2 to 6. That of 6 is stored. We place then in 1 four gallons of vinegar and 6 quarts 80% spirits, and then add water to the 35 gallon mark, to make the first wash, which is suffered to run through A. In making the different washes, a portion of the water is taken sufficiently hot to bring the wash to the proper temperature. The vinegar of 2 is then removed to mixing tub 3, and that of 4 to tub 5. To No. 3, $2\frac{1}{2}$ quarts, and to No. 5, $1\frac{1}{2}$ quarts of 80% spirits are added, and the washes are set to running. It is better, however, to add these portions of spirits differently, thus: At the commencement of running the 1st, 2d, and 3d washes, (the tubs 2, 4, and 6 being empty,) we place $2\frac{1}{2}$ quarts of spirits in No. 2, and $1\frac{1}{2}$ quarts in No. 4. The liquid which then drops from A, is forming the 2d wash in No. 2, which is made as soon as No. 2 is filled to the 35 gallon mark, when it may be transferred to tub No. 3. In like manner the 3d wash is forming in tub No. 4. A little alcohol may be evaporated in this manner; but what remains is kept longer in contact with the vinegar, and besides there is no occasion to delay the manufacture until the washes have all run from

1, 3, and 5. If the flow in one generator should lag a little, the process may go on, and the time be made up subsequently.

No. 6 may be a supernumerary tub for mixing the 1st wash, in which case a vinegar barrel, ready for storage, is placed under the third generator.

During the operation the vinegar maker tests the product to ascertain its quality. This is especially necessary at the commencement of the manufacture, when the generators are going into action. If Balling's vinegar tester be employed, it is applied to the washes in 1, 3, and 5, and the results are compared with the *vinegar* flowing from A, B, and C; that is *before* the liquid flowing from A and B is treated with fresh portions of spirits to make the 2d and 3d washes. If Otto's acetometer be used, it is applied only to the *vinegar* flowing from the three generators.

The generators may also be worked *singly, i. e.* the wash passed *three times* through the *same* generator, receiving before its passage its proper portion of alcohol. By this plan the advantage arising from keeping a body of warm vinegar in the generators is lost, since all of their liquid must be drawn before the dose of alcohol for the next wash can be added. Each generator is, as it were, set in action afresh after each withdrawal of manufactured vinegar. The method, however, permits a small business to be carried on with one generator. Two pair of large mixing tubs are required. After the last portion of each wash is added, a delay of an hour takes place to permit all of the vinegar to drain from the shavings. The proper amount of spirits is then added to the vinegar to prepare the subsequent wash. The second pair of mixing tubs is supplementary, and employed, when operating with three generators worked singly, to prevent delay arising from a lagging flow in one generator.

Points to be observed in the quick vinegar process.—Having thus given a practical example of the process, let us now consider what things are to be observed to simplify the work, to tell when it is progressing properly, and to meet and remove difficulties. The best test of a correct working is found in the temperature of the air leaving the generators. It should be uniform, and in the neighborhood of 100° Fah. It is somewhat higher during the passage of the first wash, and a little lower during that of the 2d and 3d washes. If the temperature fall much below this point, the fermentation has ceased, and immediate steps are to be taken for its restoration, by pouring on a wash heated to 100° and letting the generator stand for a short time with its top ventilating hole closed. If, on the other hand, the escaping air possesses a higher temperature, the fermentive process is indeed quickened, but the vinegar is of inferior quality, is full of vinegar eels, and of inferior strength; for not only is alcohol lost by evaporation, but a portion of acetic acid is decomposed. In order to carry on the quick vinegar method to the best advantage, a constant supervision is to be exercised over the process. Too much, generally, is expected from it. It requires much more care and watchfulness than the old method, in which nothing more was needed than regulating the temperature of the vinegar room, and giving the fermenting casks an occasional inspection. By the quick processes a constant attention must be bestowed upon many minor circumstances, upon which its successful operation depends. It is very easy to obtain vinegar of irregular acid strength, and with a considerable and variable loss of alcohol. It is by no means difficult to manufacture a first-rate article with the least waste; but method, workmen to be depended upon, and the attention to certain points, are of paramount importance. These points are not difficult to comprehend, and must be learned either by one's own experience or by that of others. I will set them forth in the following pages, mostly condensed from Otto's work on vinegar.

The quantity of air entering the generators, and the changes which it experiences.—This subject is of important consideration. The vinegar generator is a stove, in which the heat is maintained by the union of alcohol and oxygen. In order that the process goes on well the air must leave the generators at 100° Fah. There are three things which tend to draw the air through the generators. 1st. The heat caused by the acetification renders it specifically lighter, by which it rises, cooler air taking its place. 2d. A portion of its oxygen is *liquified*, becoming with the alcohol vinegar and water. This creates a partial vacuum which is filled with fresh air. 3d. The nitrogen left is specifically lighter than air, and tends to escape, being replaced by fresh air. Air contains 20.9 per cent. of oxygen, a portion only of which is employed in the quick vinegar manufacture. Knapp found in the air escaping from the generators, 19.1 per cent. of oxygen, which leaves 1·8 per cent. of this gas used for the acetification. Otto discovered by numerous experiments that the air leaving generators in good action, contained from 14 to 16 per cent. of oxygen, equivalent to from 4.9 to 6.9 per cent. of the air's oxygen employed in the vinegar fermentation. He noted, also, that in the most successful factories the most oxygen was absorbed from the air traversing the generators. Ordinarily not more than one quarter of the oxygen of the air is absorbed. It is important to render this absorption as perfect as possible, since all needless gas traversing the generators, involves loss by evaporating a portion of the alcohol. The earlier practice erred in giving too much air to the generators. A large proportion of air admitted acts injuriously, if the vinegar room be not very warm, in cooling the shavings, thus retarding the fermentation. If, to obviate this difficulty, the room be kept very warm, fuel is wasted, alcohol evaporated, and there is risk of all the disadvantages of too high a temperature in the generators. The quantity of air needed for the process is ascertained for *each generator* by carefully watching it at the outset, and regulating the current by closing to the requisite degree the hole left for this purpose in its cover. Too little air is indicated by aldehyde in the apartment, which is recognized by its penetrating odor, and by its painful effect upon the eyes. The quantity of air admitted to the generators should be just sufficient to prevent the formation of this aldehyde. The air current passing the shavings is in a measure self-regulating, because the more active the fermentative process the warmer does the air in the generator become, and the more rapidly is it deprived of its oxygen. This draws a larger quantity of air through the generator, and as the air of the apartment is lower than 100°, the shavings are slightly cooled, and the activity of the fermentation proportionally retarded. If the wash be excluded from the generators at night, the vent-holes must be completely closed to prevent too great cooling of the shavings. Otto proposes, when the wash is poured not continuously but at intervals, to perform the following experiment. After each pouring, *close the vent-holes*, and especially all of the lower ones. It is to be expected that there will be air enough in the generators to perform the acetification. Immediately before the next pouring give free admission to the air, and so on.

The Temperature of the Vinegar-room and of the Wash.—If the generators cannot be readily kept at the proper temperature; if they are too cool; pour less mixture in a given time, make the mixture warmer, or increase the temperature of the vinegar-room. If the air escaping from the generators is generally too warm, have a cooler apartment, or cooler wash, and pour more of the wash in a given time. The relative temperature of the wash and of the vinegar-room, also the quantity of wash poured in a given time, is governed by the size of the generators generally, and individually by the manner in which the shavings are packed in each generator. The quantity of mixture passing must be such that it trickles gently through the shavings, not washing them off, but becoming gradually acidified as it nears the lower layers.

The temperature and quantity of the wash must bear such a proportion to the temperature of the vinegar-room that the generators maintain the temperature of 95°–100° Fah., as the most favorable to acetification. With the average size of apartments, and kind of generators, it is enough to keep the temperature both of the wash and the vinegar-room at 73°–77°. If the apartment be somewhat cooler, the temperature of the wash must be proportionally greater, and vice versa; so that during a part of the year the room need not be heated. On the other hand, it is requisite in some localities, at mid summer, to counteract the heat of the room by some convenient and cheap method of cooling artificially the wash.

Method of Warming the Wash.—The first wash is made by the addition of *warm water* to alcohol and vinegar. The water may be heated in any convenient way; not so with the remaining washes, which contain vinegar, and which should not be brought in contact with any metal. These washes are warmed in a water bath, that is, in glass, stoneware, porcelain, and vessels placed in a receptacle containing heated water. The following cuts illustrate this method of warming the washes.

FIG. 13.

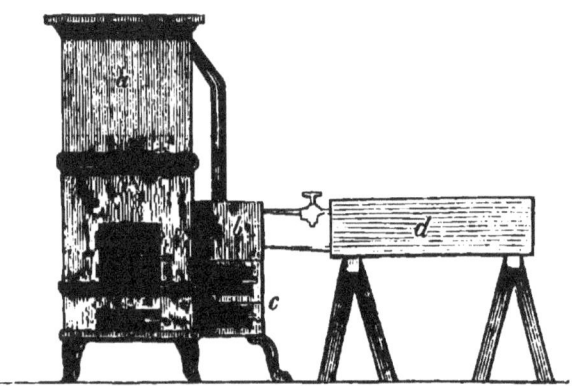

That excellent variety of stove made from earthenware or porcelain, used in Germany for warming apartments, as represented by (*a*); (*b*) is a copper vessel, one side of which is walled in the

DETAILS OF THE OPERATION. 263

stove, so that water placed in (*b*) is warmed by the fire in the stove; (*c*) is a small stove, of which the pipe ascends and joins (*a*), for the purpose of heating water when the weather is not cold enough to warm the apartment with the large stove. (*d*) is a trough, represented in figure 14, in horizontal section. It contains water, and

Fig. 14.

communicates with (*b*) by two tubes, in one of which is a cock for regulating the current. When the water in (*b*) is heated, there is a circulation in (*d*) through the two tubes; the cooler water passing from (*d*) to (*b*) by the lower tube, and the hot water from (*b*) to (*d*) by the upper tube. If the water in (*d*) gets too hot, it can be regulated by closing more or less the stop-cock which regulates the current. The trough (*d*) is divided into compartments to prevent breakage of the bottles by jarring. The longitudinal compartment enables the hot water in (*d*) to be stirred with a wooden paddle to make its temperature more uniform. Warm water for making the first wash may be drawn from (*d*), but it is better to have a separate apparatus for heating this water. Such an apparatus is represented by fig. 15, which hardly needs an explanation. It consists of a double

VINEGAR MANUFACTURE.

Fig. 15.

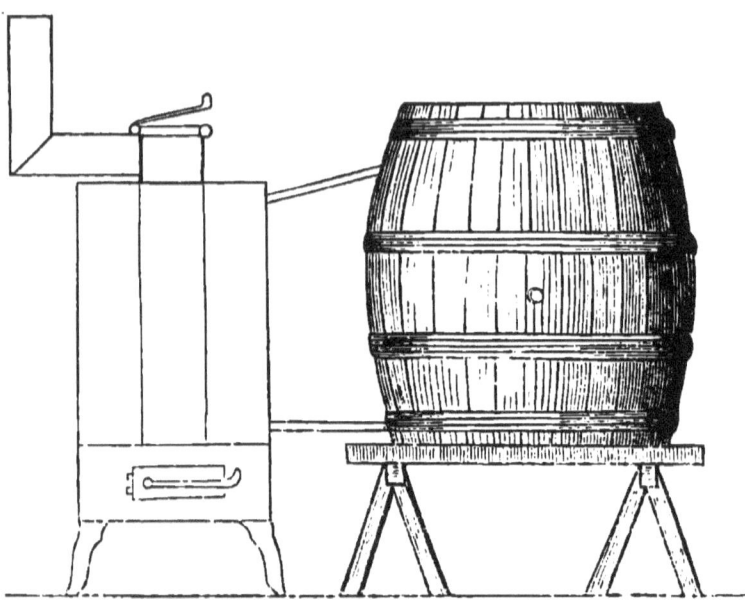

cylinder, with water between the cylinders, and heated by a fire of charcoal, wood, or stone coal, placed in the inside of the inner cylinder. This water compartment connects with a large barrel of water by an upper and lower tube, as in the cut, and a circulation of hot water is thereby effected, by which means the water in the barrel can be speedily brought to the boiling point. This heater is the only kind needed when working the generators by Otto's practice, which does not require any of the acid washes to be heated.

Otto's Practice to Avoid Heating the Acid Washes. This method requires but one mixing tub, and obliges a periodical pouring of the washes. The mixing tub must be covered with coarse paper pasted upon its outside, to enable it to retain its heat, and the first wash is placed in it sufficiently warm to maintain a proper temperature in the first generators.

Let us suppose the same example as on page 253, where the first wash consisted of $28\frac{1}{2}$ gallons of water, 4 of vinegar, and 6 quarts of 80% alcohol, where the second and third washes were made from the vinegar which had run through the first and second generators, with the addition, respectively, of $2\frac{1}{2}$ and $1\frac{1}{2}$ quarts of 80% alcohol, and where the washes ran through the shavings at the rate of $2\frac{1}{2}$ gallons per hour. By Otto's practice the same proportions are retained, but $2\frac{1}{2}$ gallons of each wash are made every hour, and poured at once on the sieve tubs of the respective generators A wooden measure, containing $2\frac{1}{2}$ gallons, is procured, and a glass measure, capable of measuring a quart and its fractional parts, fluid ounces and fluid drachms. Now by the former example, the wash ran at the rate of $2\frac{1}{2}$ gallons per hour for 14 hours. Therefore, in making a wash of the same proportions, hourly, we must take $\frac{1}{14}$ part of the whole quantity of the ingredients employed in the former instance. For example, every hour we take $\frac{1}{14}$ part of the 6 quarts of 80% alcohol, 13 fluid ounces, $5\frac{3}{4}$ fluid drachms, to which we add $\frac{1}{14}$ of 4 gallons = 1 quart $4\frac{1}{2}$ fluid ounces of vinegar, and add to these enough water to make $2\frac{1}{2}$ gallons altogether. Hence we place in the $2\frac{1}{2}$ gallon wooden measure 1 quart 4½ fluid ounces, of vinegar, and 13 fluid ounces 5¾ fluid drachms of 80% alcohol, and fill to the mark with water to make the first wash, which is all poured evenly over the sieve tub of the first generator.

When this is done we place in the wooden measure 5 fluid ounces, 5¾ fluid drachms, (that is $\frac{1}{14}$ of 2½ quarts) of 80% alcohol, and *draw from the first generator* vinegar enough to fill the measure to the 2½ gallon mark. This is the second wash, all of which is at once poured into the sieve-tub of the second generator. We then place 3 fluid ounces 3½ fluid drachms of 80° per cent. alcohol, (¼ part of 1½ quarts) in the wooden measure, and fill to the mark with vinegar from generator No. 2, to make the third wash, which is at once poured upon the sieve-tub of the third generator. To save time, a glass measure may be graduated very carefully for these three quantities of alcohol; or we may have measures for each, made of bottles of the proper size.

As will be perceived in Otto's practice, the hot vinegar of the generators is employed to make the washes, thereby avoiding the necessity of warming the second and third washes. Care must be taken to arrange the goose-neck tubes in the first and second generators, so that the air-space above the liquid shall be sufficiently large to permit it to cool down to the temperature of 73° to 77°. The temperature of the apartment may be so regulated as to obtain the required cooling. By this method the vinegar is not suffered to run in a continuous stream from any goose-neck, except that of the last generator which delivers vinegar ready for storage.

To regulate the quantity of Ferment in the Generators.—Schulze observes with respect to the vinegar *ferment*, that it brings about the decomposition of the alcohol; while the *acetic acid* determines that said decomposition shall be into acetic acid and water, and not into other products. Of what this ferment exactly consists we are yet ignorant; but that it is generated from certain nitrogenized products present in the vinegar is very probable. Some, indeed, deny the idea of a "*ferment*," properly so called, and point to the acetification of alcohol by platinum, for their reason. If this be a correct view, what the others call a "ferment," the latter characterize as a "*catalytic agent.*"*

The shavings in a generator may contain too little ferment or catalytic body, in which case the acetification takes place imperfectly, and in consequence, the apparatus maintains its heat with difficulty.

On the other hand they may be coated with too much ferment, in which case the vinegar made is proportionally *weaker*, full of eels, yellowish, cloudy, and the generators exhibit a very high

* A catalytic agent is one that produces a chemical effect by its *presence*, itself undergoing no change, like platinum in the example cited.

temperature. If the indications show too little ferment, the remedy consists in pouring lukewarm vinegar over and over again at considerable intervals, taking care that the temperature in the generators be maintained at about 106° F.

Ferment is thus generated afresh. The wash, (at first weak in alcohol) is now added, and the generators are thus brought quickly into action.

The simplest method of curing a generator suffering from excess of ferment, consists in pouring at once several buckets full of very *clear*, strong, lukewarm vinegar, which washes the excess of ferment from the shavings.

A general tendency in a generator to form excess of ferment, may be corrected by the *judicious* use of ethereal oils or antiseptics which tend to weaken the fermentative act. Care must be taken not to arrest it altogether by the use of these substances. Where the manufacture is periodical and interrupted at night, a few drops of oil of cloves, or a little alcoholic mixture which has stood over cloves, may be added to the last wash. This effectually prevents the generation of too much ferment, which is especially apt to take place at night when the shavings are saturated with a wash at rest.

The advantage of a Periodical over a Constant flow of the Wash.—I have given examples of two methods of suffering the wash to flow. In one of them the wash either runs uninterruptedly day and night, or perhaps during 12 or 14 hours resting at night; in the other a certain quantity of the wash is poured every hour. The latter practice has been almost universally adopted in Europe, and excludes many of the difficulties under which the quick process has hitherto labored. In our country the rule is, "*constant flow;*" but the time will come when, urged by competition, manufacturers will be obliged either to resort to a periodical flow of the wash, or to invent means for overcoming the objections attending a constant one. The first trouble of the inexperienced vinegar-maker, generally arises from difficulties arising from the constant flow. The sieve-tub or platform at the top of the generator is intended, like the rose of a watering-pot, to diffuse the wash equally over the shavings. For a periodical flow it contains small holes simply, and the wash is poured all over it. For a constant flow it contains four large tubes to act as chimneys, and the holes are filled with twine along which the wash trickles. In this case the wash stands at a depth of several inches upon the sieve bottom. Instead of twine, splinters of wood have been employed in the holes; also the heads of threshed rye or wheat with a couple of inches of the straw, put through the holes, the head of the rye or wheat preventing the straw from falling through. A variety of other substances have been used to replace the twine, but without any advantage, for some of the holes soon cease delivering the wash by the accumulation of "mother" on the twine. It is difficult to place the sieve-tub so that it shall remain horizontal, if it warp, the wash will flow faster through some of the holes and will not be diffused equally. It is also difficult to arrange the proportions of the holes and twine, so that the required amount of wash drops on the shavings. The best way of arranging for a constant flow, consists in boring the holes with a pod auger, then, after soaking the tub well so as to swell it, to run a red hot iron through each hole. The sieve-tub is then again soaked as well as the pieces of twine, first in hot water, and then in vinegar. The twine is chosen of such thickness that the wash covers the bottom of the sieve apparatus, and drops in a gentle shower from the twine. The flow is easily regulated by graduating the tub delivering the wash, suffering the same to flow upon the sieve through a faucet, and observing that the proper number of gallons flows per hour. It is not difficult to attain these objects for a few days, after which the "mother" begins to collect upon the twine, lessening the amount of wash that can pass in a given time; more wash flows upon one side of the generator, and at last the "mother" stops the flow effectually. The twine may be withdrawn, cleansed, and re-inserted, but the difficulty soon recurs. The following plan has been tried: A strong hoop is placed upon the cleets instead of the sieve-tub. Upon the hoop is tightly stretched a strong net having apertures of an inch. This net supports three or four little wooden tubs, communicating with each other by glass or vulcanized rubber tubes, and filled from the wash reservoirs by a tube regulated by a wooden faucet. Pieces of lampwick are so arranged, that one end of a piece is immersed in the vinegar of one of the little tubs, while the other end hangs through a mesh of the net; taking care that there is a wick for every mesh. The wash is drawn over by capillary attraction, and drops by a syphon-like action from every wick. The flow is at first admirable, but at length the capillarity of the wicks is destroyed by the accumulation of "mother." This result may be put off for awhile by placing a few cloves in each little tub; but it arrives eventually.

It would be of great advantage, if a proper constant flow of the wash could be effected. The wash placed over the generators could be warmed by the heat escaping from them; it could be employed in a cooler state than by the periodical flow, as the drops would be warmed in their passage through the upper layers of shavings. This practice would lessen the labor of attending the generators, and would permit the manufacture to go on all night without supervision. I think that a constant flow, free from the objections to it as practiced in our country, can be effected with a little ingenuity, especially in this age of vulcanized India-rubber.

A large factory in England, employing very

large generators, through which the air is *driven* by steam power, gives the hint upon which to experiment. In this factory the wash flows through a regulating stop-cock, from a reservoir placed above the generator, and enters a tube stopped at the ends, and revolving horizontally by steam power. This tube is pierced with small holes from end to end, and in its revolution drops the wash uniformly over the shavings. It would seem that a similar tube of hard vulcanized rubber, revolving by clock-work might afford a ready and constant flow of wash to an ordinary generator. The weights of the clock-work might be so arranged as to be wound up every 12 hours; which labor is certainly not equivalent to an hourly pouring of the wash, while it enables the manufacture to go on uninterruptedly during the night.

The revolving apparatus should, together with tubes and faucet, be visited from time to time, to remove the accumulated " mother." Until such an apparatus has been invented, I would advise decidedly the adoption of the German practice of a periodical pouring of the wash. Its good results are no longer problematical, the matter having been fully investigated by the most experienced vinegar-makers of Europe.

CHAPTER IV.

EXAMPLES OF THE PRACTICE OF THE BEST EUROPEAN VINEGAR FACTORIES.

OTTO has given, in his treatise on vinegar, the following valuable information respecting the actual practice in the factories of his country. In translating and condensing, I have avoided re-calculating the foreign measures into those employed by us, in order to obviate the slight error arising from neglecting fractions. It will therefore be necessary to give the value of the German measures. In the *present* chapter, the word "*quart*" (qt.—qts.) refers to the Prussian measure of that name. *Qr. Qrs.* denotes the Brunswick "*Quarter.*"

TABLE.

1 Prussian oxhoft equals 180 Prussian quarts = 206·1 litres.
1 Prussian quart . . . . . = 1·145 "
1 Brunswick oxhoft equals 240 Brunswick qrs. = 224·5 "
1 Brunswick quarter . . . . = 0·9354 "
11 Brunswick quarters equal 9 Prussian qts.
1 English quart imperial . . . = 1·136 "
1 Wine measure (United States) quart . = 0·947 "
1 quart imperial equals 1 qt.+6 oz.+3 dr.+
    16 minims wine measure.
1 United States vinegar barrel of 40 gallons
    wine measure . . . . = 151·5 "

In this table I have compared the German with

our measures, by the medium of the French litre.*
It will be seen that:

1. The Prussian quart differs but little from the Imperial quart.
2. The quart of the United States (wine and apothecaries measure,) is nearly as large as the Brunswick quarter.
3. The Prussian oxhoft contains 14 gallons 3 pints (wine measure) more than the United States vinegar barrel of 40 gallons.
4. That the Brunswick oxhoft contains 19 gallons and 1 qt. more than the United States vinegar barrel. In other words, the Prussian oxhoft equals 54 gallons + 3 pints, and the Brunswick oxhoft = 59 gallons + 1 quart wine measure.

VINEGAR PROCESS IN GERMANY.

Scarcely two vinegar factories work precisely alike, which proves that the quick process has not yet attained the desirable degree of perfection; for otherwise there would be one and the best method. The following are examples of the process as carried on in Germany.

1. *A factory in the Dutchy of Brunswick* employs 3 generators 12 feet high, and six mixing tubs for the wash. A lattice work, inside, above the lower ventilation holes, supports the shavings. A sieve platform, with holes without twine, serves to scatter the wash. The cover is air-tight, and contains a funnel for pouring the wash upon the

* 1 litre = 2·1135 pints wine measure.

sieve. A *vent tube* in this cover conducts the air from the generators *out* of the vinegar room. The generators have goose-neck tubes for the flow of the vinegar. Temperature of vinegar room 82° Fah. The flow is periodical. The first wash is prepared by placing in a mixing tub of 1 Brunswick oxhoft (that is 240 quarters) capacity, $15\frac{1}{2}$ quarters of 80% alcohol; 12 qrs. of vinegar; $1\frac{1}{2}$ pounds of *syrup;* and enough water to make up the 240 qrs. This water is taken sufficiently warm to give a wash of 100° Fah.

Every hour 8 qrs. of this mixture are poured upon generator A, and run (as four per cent. vinegar) into the second mixing tub of B, in which 9 quarters of alcohol have been placed. As soon as enough vinegar flows from A into this tub to make up 240 quarters, it becomes the 2d wash, and is poured through generator B, at the rate of 8 quarters hourly. It runs from B (as 6 per cent. vinegar) into the mixing tub of C, in which 7 quarters of alcohol have been placed. As soon as this tub is filled with 240 quarters, it becomes the 3d wash which is poured, 8 quarters hourly, upon generator C. What flows from C is saleable vinegar of 8 per cent. acid strength. As has been seen, the washes are running from three of the mixing tubs, while the next washes are being made in the remaining three mixing tubs.

2. *Factory in the city of Brunswick.*—This factory employs two generators ten feet high, which have very small vent holes.

The first wash consists of 20 quarters of vinegar; 15 of 80% alcohol; a little perfectly clear white beer; and sufficient water of 100° Fah. to make up 240 qrs.

Every 3 hours 14 qrs. are poured upon the shavings of A. Every 3 hours 14 qrs. of vinegar are drawn from A, $\frac{1}{2}$ qr. of alcohol added, and the 2d wash thus formed is poured upon B. What flows from B is manufactured vinegar. It is placed in a cask filled with beech shavings, standing in the vinegar room. It clarifies in this cask, from which it is drawn for storage.

Temperature of vinegar room, 73°—79° Fah. That of generators 100°—102° Fah.

3. *A factory in Beuthen* works with 4 generators, which are between 7—8 feet in height.

The 1st mixture contains 14 quarts (Prussian) of 80% alcohol, 50—55 qts. beer, 20 qts. vinegar, and 110 qts. water.

The 2d mixture contains 16 qts. alcohol + 14 water. Every two hours, say at 5, 7, 9, &c., o'clock, six qts. of the 1st mixture are poured upon generator A. The six qts. which have run from A, are poured upon B. To the six qts. which have run from B, $\frac{3}{4}$ qt. of mixture No. 2 are added, and poured upon C. To the 6 quarts flowing from C, $\frac{1}{4}$ qt. of mixture No. 2 is added, and the wash thus made is poured upon the last generator, D. From D, every two hours, six quarts of made vinegar are drawn.

At the intervening hours, (6, 8, 10, &c., o'clock,)

10 quarts of vinegar are drawn from generator A, and poured upon B ; and 10 qts. drawn from B, are poured upon A. Ten quarts are also poured from C to D, and from D to C.

The temperature of the vinegar room is 73° Fah. That of the 1st mixture the same.

That of the vinegar poured from the generators from 78°—86°. Consequently the generators B, C, and D, keep themselves warmer than A. The quantity of alcohol employed is considerable, and equivalent to a wash of over 12 per cent. alcoholic strength.

4. *Schulze's Method.*—Schulze, a very experienced manufacturer, obtains, by the following method, a vinegar of which an ounce requires 50 grains of carbonate of potassa for saturation, which is equal to an acid strength of 7·7 anhydrous acetic acid. He employs 3 generators 9 feet high, and two mixtures; a weak one containing 6% alcohol for A, and a strong one containing 20% alcohol for B and C. These mixtures are simply alcohol and water.

Each generator is furnished with a goose neck tube, from which flows *cooler* vinegar from the bottom of the generator ; and a faucet placed two inches above the bottom for drawing off *warmer* vinegar.

Every day, from 5 A. M. to 9 P. M., 10 Prussian quarts are poured hourly from wooden buckets through each generator as follows.

A receives 5 qts. weak mixture and 5 qts. vinegar from A. The mixture is first placed in the bucket, which is then filled to the 10 qts. mark with vinegar from A.

B is charged as follows: The vinegar which has run through the goose-neck of A (in quantity about 5 qts.) falls into a 10 qt. bucket, receives 1 qt. *strong* mixture, and is then filled to the 10 qt. mark with vinegar from B. The wash thus made is poured upon B.

C is fed in a similar manner. To the vinegar flowing from B, $\frac{1}{2}$ qt. of strong mixture is added, and then enough vinegar is drawn from C to make 10 qts. of wash, which are poured upon C. The vinegar which flows from the goose-neck of C, amounting to about 6 qts. per hour, is ready for sale.

Schulze *draws the air* through his generators from above *downward* in a manner to be described directly. The temperature in A. is from 89°—93° Fah.; in B. and C. from 84°—91°. The heat is uniform in each generator, while by the ordinary method of serving the air, the lower portions of generators are cooler by reason of the upward current of cold air.

5. *A highly prized recipe* recommends for vinegar requiring from 60 to 70 grains carbonate of potassa to saturate two ounces, (*i. e.*, 4·6—5·4 acid strength,) *four* generators, from 6 to 7 feet high, and about $3\frac{1}{2}$ feet in diameter. Also for vinegar requiring from 80 to 90 grains of the same salt for saturation, (*i. e.*, 6·1 to 7 per cent. acid,) the same number of generators, but from 8 to 9 feet high and $4\frac{1}{2}$ feet in diameter.

For vinegar of the former strength, the process is carried on as follows. Two mixtures are made; one consisting of 13 qts. 80 per cent. alcohol and 167 qts. of water. The second mixture is a certain fermented liquid called technically "*the ferment*," and is employed in making the different washes. This "*ferment*" is prepared in the following manner: forty qts. of boiling water having been placed in a cask, there are added 3 pounds purified tartar, 12 ounces tartaric acid, 3 pounds sugar, 1 pound honey, 40 qts. of beer and the same quantity of vinegar; six lemons sliced, 12 pounds of berries of the Mountain Ash, 10 qts. of diluted wine or cider, and a few cups of yeast. As soon as fermentation has set in, add daily for every 4 qts. taken away, 4 qts. of water and $\frac{1}{4}$ qt. of alcohol; every third day add 4 qts. of beer and $\frac{1}{2}$ qt. of cider, until the cask is full. The ferment is ready for use in a few days. The cask has one faucet at the bottom for emptying its contents, and another in the middle for drawing ferment for the washes.

When ferment is drawn, it is replaced by as much water, alcohol, beer and saccharine matter.

Otto remarks upon this complicated ferment, that a cask containing young white beer or malt wine, vinegar and a little spirits, will answer a better purpose, taking care to replace what is withdrawn by diluted alcohol, fresh beer, syrup and the like.

To return to the vinegar recipe. The washes are poured every hour, as follows: Upon A, $10\frac{1}{4}$ qts. of first mixture, + $\frac{3}{4}$ qt. ferment. Upon B,

11 qts. from A, + $\frac{3}{16}$ qt. each of alcohol, ferment and water. Upon C, 11 qts. from B, + $\frac{1}{8}$ qt. respectively of alcohol, ferment and water. Upon D 11 qts. from C, and $\frac{1}{16}$ qt. each of alcohol, ferment and water. From D the manufactured vinegar is drawn.

By this recipe, for 180 qts., 18 qts. of alcohol and the same quantity of ferment are employed, which is a large proportion of alcohol to make vinegar of the required strength. Blackboards are placed upon the generators for the purpose of chalking their numbers, for they are worked in rotation. Thus, what is A to-day, is B to-morrow, and D becomes A. The third day, C begins; the fourth day D, and the fifth day the series re-commences with A as the first generator.

The first generator receives the most air, the last generator the least air.

If the fermentation slackens, $\frac{1}{2}$ the quantity is poured hourly, a larger proportion of alcohol and ferment is added to the wash, and the lower ventilation holes are opened widely. If the generators become too hot, colder washes and containing less alcohol are poured. Beside, the lower vent holes are opened less than the upper ones. As said before, stronger vinegar is made with larger generators, and with more alcohol.

The generators in this recipe are constructed in a peculiar manner; they contain a lower partition, pierced with $\frac{1}{2}$ inch holes, and covered with felt, cloth, or linen. Upon this rests a three inch layer of washed gravel mixed with charcoal, upon which lies another partition pierced with holes; this arrangement constitutes a filter. Between 2½ and 3 inches above the top of the filter, a wooden air tube inclining gently upward, penetrates the side of the generator and reaches its centre. The lower side of this air tube in the generator is pierced with holes. The tube terminates *outside* in a funnel of orifice pointing downward. Upon the air tube, and upon blocks resting upon the top of the filter are willow baskets, one placed above the other for receiving the beech shavings, etc.

These baskets are loosely woven; of the diameter of the generators, and from 2 to 2½ feet in height. They are packed carefully with tightly curled beech shavings, with lumps of charcoal or with the felt shavings or cuttings of hat makers. The bottom edge of each basket is furnished with a strip of felt, fitting closely to it and to the walls of the generator to prevent an ascending air current at that place. An upper partition, pierced with small holes and covered with felt, rests a little above the top basket; immediately below this upper partition the generator is pierced with two holes, one for the thermometer, the other for an air pipe; this air pipe is of 2½ inches diameter in the clear, and is of elbow form; one branch penetrating the generator to the centre, the other rising outside vertically one foot above the cover; this branch has a valve of the kind sometimes used in stove pipes, to enable the regulation of the air current.

6. *A factory in N.* makes vinegar from 10 per cent alcohol passing through 4 generators, 11 feet high. The first wash contains 6 per cent. alcohol; the remaining 4 per cent. of alcohol are divided between the remaining three washes. Returned pourings are sometimes practiced, i. e., the liquid from B is poured again through A, etc.

7. *A factory in S.* employs generators of only 6 feet high and containing 1000 qts. (5½ Prussian oxhoft.) The mixture contains 10 per cent. of alcohol; to every 200 qts. of mixture, 20 of vinegar and 14 of white beer are added. The vinegar contains 8·4 per cent. anhydrous acetic acid, for an ounce requires for saturation 55 grains carbonate of potassa.

8. *A Manufacturer in L.* is said to make vinegar, an ounce of which requires 68 grains of carbonate of potassa for saturation, (that is, of 10·4 acid strength,) from a mixture containing 10 per cent. alcohol.* He employs 4 generators, to which the air is admitted *below* the shavings by means of a tube in the shape of a cross, and pierced underneath with holes. One arm of the cross is prolonged through the side of the generator to the external air and terminates there in a funnel, the larger aperture of which points downward.

The mixture consists of alcohol and water; thus

* There is probably a typographical error at this place in Otto's work. The wash which follows contains only 9% absolute alcohol by volume, which by Table II, p. 204, cannot yield a vinegar containing more than 7·6% alcohol

for the first wash, 15 qts. of 80% alcohol with 192 qts. of water. The vinegar from this passes the second generator without further addition of alcohol; 9½ qts. of alcohol are divided between the third and fourth washes.

9. *In some factories* a mixture consisting of alcohol and water, with some white beer, malt wine, cider, etc., is poured hourly upon the generators and returned until the vinegar has acquired sufficient strength. The lower vent holes are placed at fifteen inches above the bottom of the generators, in order to always keep a body of vinegar in them. For generators of from 6 to 8 feet high, 12 qts. of wash are poured hourly, and the pouring is frequently crossed; *i. e.*, the vinegar from B is poured upon A, and that of A upon B, etc. The temperature of the vinegar room is kept between 68°—77° Fah. In the generators, it is from 91°—100°, and rising at night to above 100°. This is in effect a modified Boerhave process.

The following, still more resembling the method of Boerhave, is also practiced. Very wide generators, without vent holes, and with tight covers, through which air may be admitted at intervals, are employed. The liquid is drawn from the generators, and poured every twelve to twenty-four hours. Other factories employ similar generators, but with lower vent holes, which are open three or four times a day.

10. *Improvements upon the quick vinegar process in England and Germany.*—In England large vinegar factories have come into practice, and

improvements have been introduced in certain details of the process. These Trenn and Schulze have modified and brought into use here and there in Germany. The English generators are very large, having for a height of thirteen feet, a lower diameter of fourteen and an upper diameter of fifteen feet, equal to a capacity of 2145 cubic feet. Two and a half feet above the bottom is a platform pierced with holes. The filling consists of small blocks, cooper's shavings or chips, which reach almost to the cover. At a moderate distance above the cover are situated the wash reservoirs, which deliver their contents by a vertical tube descending through the cover, where it terminates in a horizontal tube, pierced and revolving by steam power. The air is delivered to the generators by means of two floating gasometers, which alternately rise and fall, operated by steam machinery. As each gasometer ascends, it draws its air by a pipe from the space under the false bottom of the generator. In falling, it delivers this air by another pipe into a cistern of water, thereby condensing the alcoholic vapors it contains. This water is used in making the washes. The fresh air is admitted through the cover of the generator by the side of the wash pipe, and proceeds downwards. A small forcing pump is continually raising the liquor from the bottom of the generator to the reservoir above. The acetification is governed by regulating the warmth of the apartment, and the motion of the gasometers which furnish the air. In a few days a merchantable vinegar of 4·7% anhydrous acid is obtained. One of these large generators has the same power as six of eight feet high and four feet diameter. Knapp has set forth the comparative advantages of the large over the small generators as follows:

1st. On account of the greater capacity of the former, their elevated temperature may be more readily maintained, which results in such a saving of heat that fires may be dispensed with in mild winters. A large generator, as described, contains 611 square feet of *stave* surface. Six smaller generators of the above mentioned proportions, contain 603 square feet of stave surface. The surfaces are not very different, but the *cubic contents* are in the ratio of 2287 cubic feet to 603, that is as 3·79 : 1.

2d. Small, irregularly shaped blocks or chips distribute the wash more uniformly, and since, on account of their incompressibility, they do not change position by time like shavings, the equable diffusion of the wash is lasting. In some parts of Germany they employ cubical blocks bored with three large holes from face to face of the cube.

3d. The diffusion of the wash over the blocks is more uniform by the English than in the periodical flow, and it is uninterrupted.

4th. The admission of air is independent of the temperature of the generator, and may be altered to suit the nature of the wash and the strength of its flow.

5th. The loss of alcohol by evaporation is almost nothing.

Trenn in Germany, took out a patent for drawing a downward current of air through the generators by means of a furnace. Schulze improved upon this process as follows:

The generators are nine feet high and three feet in diameter. Larger ones become too warm even when the temperature of the apartment is very low, and involve thereby a loss of alcohol by evaporation. At a distance of ten inches above the bottom, a latticed platform rests upon two oak cross pieces. An air tube turned of wood, of one and a half inch bore, penetrates the bottom of each generator, reaching nearly to the false bottom, and furnished at the end with a little roof to prevent intrusion of drops of vinegar. The generators are furnished with goose neck tubes of glass or gutta-percha, and with a faucet, which is placed at two inches above the bottom.

The filling is clean charcoal from soft wood. The lumps of inferior strata are of the size of walnuts; the higher ones of hazel nut size; and the top layers as small as peas. Placed upon this stratum is the sieve partition, furnished with four glass chimney tubes. The coal is removed immediately under these tubes, leaving four small cavities. The cover of the generator is perfectly airtight, and contains a shutter ten inches square, for the periodical admission of the wash. In this shutter is a vent hole of one and a quarter inches in diameter. The peculiarity of this process consists in the apparatus for feeding the generators with a downward current of air.

If a suction be effected at the vent tube in the bottom of the generator, air will be drawn from the upper part of the vinegar room through the vent hole in the cover of the generator. This suction is obtained by means of a furnace represented by Figure 16.

Fig 16.

As may be seen, this furnace is constructed with a double door, and with very thick walls, and close ash pit door, to retain as much heat as possible. The flue from the fire place, (represented in section at d,) is *wide* but flat, that is, with two of its walls close together. It communicates with a chimney by means of another flue (c), which rises three inches in the foot. In this inclined flue are placed, side by side, and two and half inches apart, cast-iron tubes ($a\ b$), of one and a quarter inch bore, three-eighth inch thickness of metal, and five feet long. There is such a pipe for every generator. Each tube has a direct communication (at a), by means of a wooden tube covered with a bad conductor of heat, with one of the generators. These wooden tubes proceed from a down to the floor, and rise to unite with the respective vent tubes in the bottom of the generators. A small fire is made in the furnace only once in twenty-four hours; for, by reason of the thickness of the walls, the heat is retained in the furnace, and the cast-iron tubes are kept warm and continue sucking air through the generators all day. The cost for fuel is no greater than for the ordinary way of working generators, because the excess in summer is balanced by the saving in winter. The air being drawn from the top of the apartment where it is warmer, saves a portion of the fuel required for heating the vinegar room in the winter time.

CHAPTER V.

CONCLUSION.

In concluding this work, let us note several subjects worthy of the attention of the vinegar manufacturer.

FLAVOR AND ODOR.

Vinegar, made from pure alcohol and water, does not possess the pleasant aroma of wine or cider vinegar, and is therefore inferior to them for table use. Modern discoveries, especially those of the volatile ethers, enable us to overcome this objection in a great measure, and it is to be expected that future experiments will remove it entirely. Some of these ethers are sold at a very high price, because they are kept secret for a particular use. This will not exclude their employment by the vinegar manufacturer, for, if available, they may be made by cheaper methods. Many aromatic substances added to the wash, are completely changed in the quick process. There is a case on record where camphor had been dissolved in alcohol to defraud the revenue. This alcohol yielded by the quick process a vinegar without the slightest smell of camphor.

The disagreeably smelling fusel oil, which exists in raw spirits obtained from rye, maize, wheat, &c., is changed by the acetification into an agreeably smelling ether.

Potato fusel oil gives a less pleasantly flavored product. Hence raw spirits give a product resembling cider vinegar more than that from rectified spirits. And hence the addition to rectified spirits of a few drops of fusel oil, in making the wash for the vinegar process is of advantage.

A little added oil of cloves, or extract of this spice gives, after acetification, a still more pleasantly flavored vinegar. The addition of a little butyric ether, valerianic ether, or one of the vegetable vinegars (to be described directly) to the last wash, yields a vinegar of very agreeable flavor. There is room here for experiment.

To give it flavor, vinegar is sometimes stored in casks containing powdered tartar, crushed raisins, and raisin stems. A few drops of a mixture of 10 parts acetic ether, and 1 part pear ether (valerianate of amyle); or the addition of a small quantity of vegetable vinegar, gives to a large quantity of vinegar a very agreeable flavor. These additions may also be made to the last wash.

Finally, a portion of the last wash may be wine or cider.

COLOR.

Since the vinegar made by the quick process is limpid as water, it is necessary, to suit the public taste, to color it. This coloring is simple, perfectly harmless, and, perhaps, not very different from that existing naturally in fruit vinegar. One method consists in roasting barley malt and using a sufficiently strong infusion made with lukewarm water.

An aqueous solution of sugar, that has been melted in a clean brass or copper kettle until it acquires a dark color, will effect the same purpose.

A very common coloring in Germany is the infusion of chicory coffee.

To imitate red wine vinegar, the juice of very ripe black mulberries may be employed.

CLEARING VINEGAR.

Clearing vessels for vinegar should be found in every well arranged factory. These may be upright casks filled with closely curled beech shavings, and with a wooden faucet at the bottom for drawing off the vinegar after it has stood for some time upon the shavings.

The following is a convenient filter for the same purpose.

Take a cask from 2 to $2\frac{1}{2}$ feet in diameter, and from 3 to 4 feet high. Stand it on end, placing a faucet at the bottom. Charge then with a 5 to 6 inch layer of charcoal of hazel-nut size, then with a 3 to 4 inch layer of finer coal. After this is well leveled, add an inch stratum of *paper pulp*, then another layer of fine, and one of coarser coal, and fill the rest of the vessel with closely curled spirals of beech shavings. The coal must be well washed and soaked in vinegar. If a vessel be made for the purpose, it had better be somewhat conical, and placed with the greater diameter uppermost. It must be kept always full of liquid or it will act like a generator. In cleansing the filter, the paper may be used again, by placing on a fine sieve and pouring boiling water upon it. This filter yields a perfectly clear vinegar, and doubtless of better keeping qualities than if not filtered.

VEGETABLE AND AROMATIC VINEGARS.

These are fancy vinegars, generally infusions of herbs or spices, and employed for fumigation, for the toilet, or as additions to the last wash in the quick process, for the purpose of communicating an agreeable aroma to the resulting vinegar, rendering it more valuable for table use.

Four Thieves Vinegar.—Said to have been invented during a great plague in Marseilles, (some say during the London plague,) by four thieves, who employed it to prevent infection during their predatory visits to the houses of the dead or absent.

Macerate cloves, sage, rosemary, rue, allspice, calamus, caraway, nutmegs, of each one ounce, in two gallons of strong vinegar. Then add half an ounce of camphor.

Another Way.—Wormwood, rosemary, sage, peppermint, rue, each 2 ounces, lavender blossoms, 6 ounces, calamus, cinnamon, cloves, nutmegs, (garlic,) each 1 ounce. Macerate in two gallons of vinegar; then add a little tincture of camphor.

A Third Method.—Rosemary, sage, peppermint, cloves, of each 4 ounces; zedoary and angelica roots, of each 1 ounce. Macerate for several days in half a gallon of vinegar. Then press and filter.

Tarragon Vinegar.—Soak for several days 1 pound of the herb, (*Artemesia Dracunculus,*) before blossoming, in from 1 to 2 gallons of very strong vinegar. Press out the liquid and filter.

The vinegar may be made extemporaneously by dropping a few drops of the oil of tarragon upon a lump of sugar, and adding to the vinegar. For table use the vinegar should not be too strongly flavored with the herb.

Vinaigre aux fines herbes.—Tarragon, (herb,) 12 ounces; basil, (herb,) 4 ounces; laurel leaves, 4 ounces; shallots, (*Allium Ascalonicum,*) 2 ounces: are suffered to stand for a few days in $\frac{1}{2}$ a gallon of strong vinegar. Press and filter. A little added to table vinegar improves it.

Vinaigre à la Ravigote.—Tarragon, (herb,) 12 ounces; laurel leaves, 6 ounces; anchovies, 6 ounces; capers, 6 ounces; shallots, 4 ounces. Macerate for several days in $\frac{1}{2}$ gallon of strong vinegar, then press and filter. Used as addition to table vinegars.

Mustard vinegar is also employed as an addition to vinegar, and is made by soaking from 8 to 12 ounces of black mustard in 1 quart of strong vinegar. Press and filter.

Raspberry vinegar.—Ripe berries are pressed and suffered to stand for several days, after which the clear juice may be separated. To every pound of berries add from 6 to 8 quarts of strong vinegar, and press after 24 hours. Sweetened with sugar is used as an agreeable summer drink.

Rose, orange blossom, neroli, bergamot, and clove vinegars, may be made by adding the respective oils to vinegar. They may be added to the last wash in the quick process to yield a finely flavored vinegar.

Fumigating vinegar.—Oils of cloves, $1\frac{1}{3}$ drachms; bergamot, 3 drachms; cassia, 1 drachm; balsam of Peru, 2 drachms; tincture of musk, 1 drachm. Add 24 ounces of 80 per cent. alcohol, and enough concentrated acetic acid* to keep the oils in solution.

Aromatic vinegar.—Oils of cloves, 3 drachms; lavender, 2 drachms; lemon, 2 drachms; bergamot, 1 drachm; thyme, 1 drachm; cinnamon, 30 drops. Dissolved in 6 ounces of concentrated acetic acid.

Another.—Equal parts of concentrated acetic acid and acetic ether, with a few drops of oil of cloves.

Crème de vinaigre.—Oils of bergamot, 3 ounces; lemon, 2 ounces; neroli, 1 ounce; mace, $\frac{1}{4}$ of an ounce; cloves, $\frac{1}{4}$ of an ounce. Dissolve in two pounds of strong alcohol, and 5 pounds concentrated acetic acid.

* Of 25—30% acid strength.

INDEX.

Acetic acid, composition..	38
conversion of hydrated and anhydrous per cents	169, 170
derived from alcohol............................	163
density table..	166
glacial...	16
quantity obtained from alcohol..........	163
Acetometry..	172
Acetometer...	144, 167
Otto's..	189
Balling's	202
Acidity of fermented drinks..................................	25
Adulterations of vinegar..	170
Air, quantity entering generators, and its change	258
ventilation of generators by a furnace...........	287
composition of..	32
Alcohol, absolute..	127
composition of.....................................	38
rectification of.....................................	128
tests for wines and beer.......................	156
by the saccharometer.....................	157
Alcoholic solutions, properties of......................	126
strength of.......................	130
specific gravity table for........	150
transformation of volume and weight per cents..............	153
to make definite mixtures of...	154
Alcoholometer...	144
tables for temperature.....................	147, 148
mode of applying tables.................	149
degrees converted to specific gravities	194
Aldehyde...	163, 236
Ammonia solution, Otto's....................................	193
Analysis, (see tests on under subjects.)	
Animalculæ of vinegar...	212
Archimedes, principal of..	135
Areometers...	137
Aroma, of wines...	103
of Jargonelle pear....................................	24
of apple, pine apple, quince, whisky...........	24
Aromatic vinegar..	292
Attenuation of worts..	157
of real and apparent.....................	158
tables...	160, 161

INDEX.

Atomic theory	42
Apparatus for quick vinegar manufacture	238
Apparatus, (see under subjects.)	
Beaumé's hydrometer tables	141, 142
Beech shavings	241
Boerhave vinegar process	22, 233
Bouquet of wines	24, 103
Brewing, art of	106
Burettes	185
Carbon	31
Carbonic acid	32
Cellulose	51
Chemical combination, laws of	34
Clearing vinegar	291
Color of vinegar to improve	290
Compounds of nitrogen and oxygen	38, 46
Compressor	186
Couching of malt	111
Cuts, No. 1. Glass hydrometer	138
2. Litre measure	183
3. Pipette	185
4. Burette	186
5. Mohr's	186
6. Spring clip	187
7. Otto's acetometer	190
8. Plane for beech shavings	240
9. Sieve partition	243
10. Generator	244
11. Ventilating tube	245
12. Improved generator faucet	246
13. Warming apparatus for wash	262
14. Hot water trough	263
15. Warming apparatus for wash	264
16. Furnace for drawing air through generators	287
Dextrine	53, 64
Development of plants	61
Diastase	60, 109
saccharifying power of	115
Dilatometer	130
Drying of malt	113
Duration of slow vinegar process	228
Eels of vinegar	212
Equivalent	37
Factory building for vinegar	218, 225
process by slow method	223
by quick method	247
Factories, vinegar, in Europe, Part II., Chap. IV.	

INDEX.

Faucet, to keep body of vinegar in generator.........	246
Ferments................................,..............................	91
Ferment to regulate in generator.................:...	267
Fermentation, Liebig's theory of.......................	92
condition for.............................	93
wine........―...............................	99
of worts...	123
Filter for water..	217
for vinegar..	291
Flavor of vinegar to improve............................	24, 289
Furnace for ventilating generators.....................	287
Generators for vinegars....................................	243, 244
to set in action............,....................	247
mode of working three......................	254
advantage of large ones....................	285
ventilation of by a furnace.................	287
Glacial acetic acid...	16
Goose-neck tube for generators........................	244
Gums...	63
Gyle-tum..	123
History of vinegar...	13, 21
Honey added to vinegar..............................	211
Household manufacture of vinegar....................	221
Hydrogen..	33
Hydrometer...	137
, mode of using.....................	145
tables..	141, 142
for vinegar.....................................	167
Improved vinegar..	25
Knapp's calculation respecting large generators...	285
Law of proportion...	36
Laws of chemical combination..........................	34
Lamp without flame.......................................	236
Limit of vinegar manufacture.....................・.........	23
Litre measure...	183
Malt, characters of good.................................	114
quantity obtained...........................	114
extract from different grains.................	119
wine	125
Malting process..	108
Manufacture of vinegar, (see Vinegar.)	
Mashing...	115
Mash tun..	116
Measures, French..	182
German..	273
Mixtures definite of alcohol and water...............	154
Mixture, alcoholic, for vinegar.........................	213, 222,
225, 226, 248, 253, and Part II., Chap. IV.	

Modifications of slow vinegar process.................. 229, 230
       in setting generators in action.......... 249
Mother of vinegar... 212
Mohr's spring clip.. 186
Mycoderma aceti... 212
Nitrogen 33
       compounds.. 38, 46
Oxygen... 32
Orleans vinegar manufacture............................ 224
Periodical flow of the wash, advantages of........... 268
Pipette.. 185
Plane for beech shavings................................. 240
Plants, development....................................... 61
Platinum, black or sponge............................... 235
Polarized light.. 83
       action of sugar on....................... 86
       analysis of sugar by..................... 88
Proof spirits.. 145
Pyroxilic acid.. 18
Quantity of vinegar made daily by slow process..... 231
Recipes to prepare fancy vinegars..................... 292
Saccharometer 143
       may be employed for determining
       alcohol per centage................. 157
Saccharoid bodies... 49
Shavings for vinegar manufacture..................... 241
Sieve partition for vinegar generators................. 243
Slow process for vinegar................................ 220–223
Specific gravity to determine........................... 132
       as a test for alcohol.................... 132
Spring clip... 186
Starch... 54
       preparation... 57, 58
       chemical nature of.............................. 59
Steeping of malt.. 110
Storage of vinegar... 229
Sugar, composition of..................................... 38, 68
       amount used by different nations............ 67
       prepared from cellulose........................ 53
       prepared from starch........................... 80
       milk... 68
       cane... 71
       fruit... 75
       raisin... 76
       analysis of.. 79
       transformation to alcohol..................... 96
       quantity of alcohol from....................... 97
       quantity from starch and gum............... 98

INDEX.

Sugar, specific gravity of its solutions...............	121
as an addition to vinegar........................	211
Symbols, chemical...	39
Tables—saccharoid bodies.......................................	49
sweetness and acidity of wines..................	100
alcoholic contents of wines.....................	101
specific gravity of saccharometer degrees	121
Beaumé's hydrometers........................	141
corrections for temperature for the alcoholometer...	147, 148
specific gravity of alcoholic mixtures......	150
transformation of alcoholic weights and volumes per cent.............................	153
for real attenuation...............................	160
for apparent attenuation.......................	161
density of acetic acid solutions...............	166
transformation of hydrated and anhydrous per cents of acetic acid.....................	169, 170
vinegar tests.......................................	178
for alkaline carb. solutions	180, 181
alcoholometer degrees converted to specific gravities..	194
to dilute aquæ ammonia for Otto's vinegar test...	195
for determining strength of vinegar resulting from definite alcoholic mixtures	204
time required by the slow process of vinegar for acetification........................	228
German measures.............................	272
United States weights........................	197
Taste of vinegar to improve............................	25, 289
Temperature of vinegar factories and for the wash	261
Tests for alcoholic strength of solutions......	130, 131, 150, 157
for strength of saccharine solutions...........	79, 88, 121
for the acid strength of vinegar	174, 177, 178, 189, 202
for the adulterations of vinegar.................	170
for the impurities of water.......................	215
Theory, atomic...	42
of the vinegar manufacture...................	163
Ventilating tube for generators.......................	245
Vinegar, acid strength compared with alcoholic strength of the wash............................	203
to clear..	291
process depends on...........................	164
quantity resulting by the platinum process..	237
storage of.......................................	229
vegetable and aromatic......................	292

INDEX.

Vinegar manufacture,	general details of.............	209
	factory building...............	218
	slow process.....................	220-223
	in the household...............	221
	in Orleans......................	224
	quantity made per day by slow process..................	231
	the quick process.............	233
	Boerhave's method............	233
	Doebereiner's method........	234
	division of the subject.......	237
	generators...................	242, 247
	details of the operation......	247
	practical example of..........	252
	points to be observed in.....	257
	to avoid warming the acid washes......................	264
	to regulate ferment in generators......................	267
	improvements of in England and Germany...............	283
	in European factories, Part II., Chap. IV.	
Vinegar, tests of, adulteration.........................		170
of acid strength... 166, 174, 177, 178, 189, 202		
Wash, (see also mixture, alcoholic.)		
	pouring the..	251
	methods of warming...........................	262, 263
	to avoid warming..............................	264
	flow of...	268
	to effect improved constant flow.........	271
Water for vinegar manufacture.......................		214
	tests for its purity.............................	215
	to improve......................................	216
	to filter for vinegar purposes..................	217
Weights, American..		197
	French..	182
Wines, fermentation.....................................		99
	table of sweetness and acidity...............	100
	table of alcoholic strength...................	101
	aroma of..	103
Wood vinegar...		17
Worts, preparation of...................................		115
	concentration	118-120
	boiling...	122
	fermentation...................................	123
	attenuation of.................................	157, 158
Yeast, composition of...................................		92

THE APPARATUS AND CHEMICALS MENTIONED IN THIS WORK WILL BE FURNISHED AT THE PRICES ANNEXED.

ALL THE ARTICLES WILL BE OF THE BEST CHARACTER.

Chemical Thermometer, enclosed in a glass tube, to 212° F. $1 00
Glass Jar for floating hydrometer, ... 50
Saccharometer, showing per centage of sugar, . . 1 00
Specific Gravity Bottles, 1000 gr., (not stoppered,) in tin
   case, with counterpoise weight, ... 1 25
Specific Gravity Bottles, stoppered, in case, with counter-
   poise weight, ... 3 00
Specific Gravity Bottles, stoppered, 100 gr., in case, with
   counterpoise weight, ... 2 00
Gay Lussac's Volumeters, in setts of five, each, . . 75
Litre Flask, (page 183,) each, ... 60
Specific Gravity Acetometer, (page 167,) each, . . 1 25
Vinegar Pipette, (page 185,) each, ... 75
Burette, (fig. 4, page 186,) 2 00
  " (fig. 5, page 186,) without stand, ... 2 50
  " " "   with stand, . . . 3 25
Otto's Acetometer, (page 190,) . . . $1 00 to 1 50
Hydrometer graduated between 0·951 and 0·078, (page 193,)
   for Ammonia Solution, ... 1 25
Litmus Paper, per sheet, ... 5
Pure Acetic Acid, of ten per cent., (page 198,) . . 50
  " Tartaric " (page 200,) 12
Weight of 1·47 grammes, ... 25
A neat case containing the apparatus described on page
   201, without scales, $5 00; with scales and weights
   from 4 oz. Troy, to ½ gr., ... 6 50
Ballings Vinegar Hydrometer, (page 202,) . . . 1 50
Test Tube Stand and 12 Tubes, ... 1 00
Re-agents for testing water, (page 215,) viz: Solutions of
   Carb. of Soda, Oxalic Acid, Chld. of Barium, Nitrate
   of Silver, Tinct. of Galls. In 4 oz. glass stoppered
   bottles, ... 1 00
Re-agents for testing adulterations in vinegar, (page 171,)
   viz: 12 Test Tubes and Stand, 1 Glass Spirit Lamp,
   small, Solutions of Chld. of Calcium, Chld. of Barium,
   Nit. of Baryta, Baryta Water, Arsenic Acid, Solut. of
   Indigo, Nit. of Silver, Sulphuret of Iron, Yellow Pruss.
   of Potash, Iodide of Potassium, Flask for generating
   Sulphuretted Hydrogen, 4 oz. Porcelain Erass. Dish.
   The chemicals in 4 oz. glass stoppered bottles, . . 3 75

www.ingramcontent.com/pod-product-compliance
Lightning Source LLC
Chambersburg PA
CBHW022116230426
43672CB00008B/1408